M. Suzuki R. Kubo (Eds.)

Evolutionary Trends in the Physical Sciences

Proceedings
of the Yoshio Nishina Centennial Symposium,
Tokyo, Japan, December 5–7, 1990

With 103 Figures

Springer-Verlag
Berlin Heidelberg New York
London Paris Tokyo
Hong Kong Barcelona
Budapest

Professor Dr. Masuo Suzuki
Department of Physics, Faculty of Science,
University of Tokyo, 7-3-1 Hongo, Bunkyo-ku,
Tokyo 113, Japan

Professor Dr. Ryogo Kubo
Faculty of Science and Technology, Keio University,
14-1 Hiyoshi, 3-chome, Kohokuku, Yokohama 223, Japan

ISBN 3-540-54568-9 Springer-Verlag Berlin Heidelberg New York
ISBN 0-387-54568-9 Springer-Verlag New York Berlin Heidelberg

Typesetting: Camera ready by authors

54/3140-543210 – Printed on acid-free paper

Preface

Dr. Yoshio Nishina was a key figure in modern physics in Japan and a world pioneer in many fields of modern science such as nuclear physics, cosmic-ray physics, and radiobiology. He devoted his life to the development of science, so that his beloved country could compete with any other country in science and technology. Unfortunately, he died soon after the Second World War and did not witness the results of his great efforts.

To commemorate the centennial of Dr. Nishina's birth, a Nishina Centennial Symposium was held in Tokyo from December 5 to 7, 1990, under the co-sponsorship of the Nishina Memorial Foundation and RIKEN (the Institute of Physical and Chemical Research). The symposium was entitled *Evolutionary Trends in the Physical Sciences.*

The title of the symposium was very broad and ambitious. Indeed, progress in physics over recent decades has been truly amazing, so much so that the present frontiers of physics extend far beyond the horizons we saw when we were young. Experiments in particle physics have revealed many new particles, and may eventually lead to the clarification of the ultimate structure of matter, though it is not known whether man will ever fully understand how natural forces are unified. At the same time, it is becoming more and more likely that the creation of the universe will finally be discovered by continuing the lines of research into physics that have been pursued over the past decades. The physics of complex materials has been shown to be extremely rich, probably much richer than anyone realized 20 years ago.

Chemistry and biology are taking advantage of the advanced methods of microscopic observation made possible by progress in modern physics. It should not be overlooked that the physics of complex systems is now emerging as a new science which may extend beyond the limits of traditional natural science. The use of advanced physical methods has revolutionized almost every discipline in neighboring sciences. On the other hand, the extremely rapid changes in every field are making it very difficult for researchers working in neighboring fields to communicate. This is one reason why the broad scope of this symposium was welcomed by its participants.

Organizing this symposium was not easy. Fortunately, in response to our invitations, many distinguished scientists favored us and talked on subjects of their choice or subjects suggested by us. Within the limited time of the symposium, it was rather difficult to accommodate all the proposed lectures. Further, we had to limit participation to under 400 persons because of the limited seating in the auditorium.

Therefore we decided to publish the proceedings in order to make the material accessible to those people who were interested in participating but were unable to be present. The Organizing Committee is grateful to Dr. H. Lotsch of Springer-Verlag for his cooperation in making publication possible.

Finally, but not least, we would like to express our sincere thanks to the Japan Medical Association and the Japan Radioisotope Association for their generous support, also to the Ministry of Education, Science and Culture and the Agency of Science and Technology for their moral support. Without the support of these organizations, the symposium would not have been such a success.

Tokyo
January 1991

Masuo Suzuki
Ryogo Kubo

Chairman

R. Kubo (Keio Univ. and Nishina Memorial Foundation)

Organizing Committee

J. Fujioka (RIKEN)
K. Kamata (Nishina Memorial Foundation)
H. Kamitsubo (RIKEN)
T. Marumori (Univ. of Tsukuba)
R. Nakane (Nishina Memorial Foundation)
K. Nishijima (Chuo Univ.)
J. Nishimura (Institute of Space and Astronautical Science)
Y. Nishina (Tohoku Univ.)
M. Oda (RIKEN)
T. Sata (RIKEN)
H. Sugawara (KEK)
M. Suzuki (Univ. of Tokyo)
T. Yamazaki (Univ. of Tokyo)

Program Committee

K. Kamata (Nishina Memorial Foundation)
M. Ichimura (Univ. of Tokyo)
T. Marumori (Univ. of Tsukuba)
T. Masumi (Univ. of Tokyo)
K. Nishijima (Chuo Univ.)
Y. Nishina (Tohoku Univ.)
F. Sakata (Univ. of Tokyo)
H. Sato (Kyoto Univ.)
M. Suzuki (Univ. of Tokyo)
M. Takami (RIKEN)
Y. Totsuka (Univ. of Tokyo)

Symposium Secretaries

H. Sato (RIKEN)
T. Iwaki (RIKEN)

Contents

Part I

Memorial Session

Yoshio Nishina, the Pioneer of Modern Physics in Japan

R. Kubo

Nishina Memorial Foundation, 28–45, Honkomagome, 2-chome, Bunkyo-ku, Tokyo 113–91, Japan

We are gathered here today to celebrate the centennial anniversary of Dr. Yoshio Nishina [1]. He was born one hundred years ago in 1890 in a small village called Satosho near the city of Okayama as the eighth child of a respected family. His grandfather was the local governor of that area. There still remains the old house where he spent his childhood with his parents and family. The house is now restored and is open to the public as a museum to commemorate the great man of whom the village can rightly be proud. I visited the place and was very much impressed by his notebooks, handwritings, and drawings of his schooldays. All are very beautifully done proving that he was extremely bright and was regarded as a genius.

After finishing the local elementary school and the middle school, he entered the sixth national high school at Okayama. The national high school at that time was completely different from high school of today. It was an elite school corresponding to junior college for students proceeding to imperial universities. Nishina must have enjoyed his youth there. He liked to study by himself but was also a sportsman.

After this, Nishina went to Tokyo to have his undergraduate education in the Engineering School of Tokyo Imperial University. He elected electrical engineering as his major and graduated from the Electrical Engineering Department with honors receiving a silver watch from the Emperor.

But Nishina did not want to work in industry as an electrical engineer. He wanted to do something more useful than just engineering and more attractive and worth devoting his unusual talent. He considered electrical engineering as more or less a finished discipline. He thought electrochemistry was more attractive. So he accepted an invitation from Professor Kujirai, who had just started his new laboratory at RIKEN, a newly established research institute for physics and chemistry. This institute was planned taking the Kaiserliche Institut of Germany as a model. So Nishina became a research fellow at RIKEN and concurrently he registered at the graduate school of physics of the Tokyo Imperial University to study physics under the guidance of Professor Nagaoka, who was concurrently a chief researcher at RIKEN holding a laboratory there. Professor Nagaoka was the most influential person in the scientific community of Japan at that time. After three years of studying physics, fortune smiled on him. In the spring of 1921 he was ordered by RIKEN to go to Europe to study physics. This must have been a great encouragement for him. New physics

Springer Proceedings in Physics, Volume 57
Evolutionary Trends in the Physical Sciences Eds.: M. Suzuki and R. Kubo

was just being born in Europe. Europe was boiling with expectation of new ideas and new discoveries. This lucky event matched his talent and ambition. Fate made him stay in Europe for seven years, which he himself probably did not originally plan. This transformed him into a first rate physicist, which would not have been really possible if he had remained in Japan.

Nishina went first to Cambridge because Nagaoka introduced him to Rutherford. He stayed at the Cavendish Laboratory and performed some experiments on Compton scattering, which were not so successful but gave him valuable experience for his later studies. Nishina eagerly wished to work with Niels Bohr at Copenhagen and asked Bohr if this was possible. Bohr was kind enough to accept him at his Institute. So Nishina was able to move to Copenhagen in 1923. There Nishina started working on the X-ray spectroscopy of atoms under Hevesy's guidance, and soon was able to publish his first scientific paper [2] on the X-ray absorption spectra in the L-series of the elements La(57) to Hf(72) with the co-authorship of Coster and Werner. He continued the X-ray work after both Hevesy and Coster left Copenhagen and became the leader of the X-ray spectroscopy group. Indeed, Nishina made significant contributions in this field [3]. This was from 1923 to 1926, just the period when new quantum mechanics was rapidly developing. Copenhagen was the center of the revolution. Niels Bohr was the leader at revolution surrounded by its stars. There was great excitement every day. How happy Nishina was to be at the very center of this great revolution and close to the great leader and brilliant young pioneers. This excitement is vividly seen in Nishina's notes kept in the archives of the Nishina Memorial Foundation.

However, Nishina had already stayed in Copenhagen for quite a long time. The grant from RIKEN was already discontinued. His expenses were supported by his relatives at home. Also, Bohr was kind enough to arrange a grant from the Danish Government which lasted for three years. It was about time for Nishina to return home. But he wanted to do something significant in theoretical physics which he could bring back to Japan after staying so many years at the center of theoretical physics of the world. Indeed, Nishina had been interested in theoretical physics ever since the time he turned to physics. Before coming to Copenhagen, he attended in Goettingen the lectures by Born and Hilbert. Now he wished to study theory more seriously. So he went to Hamburg in February of 1928 to study under Pauli and there he worked out a theoretical paper in collaboration with Rabi [4].

Coming back to Copenhagen he decided to take up a new problem. He thus chose the theory of Compton scattering for which he had kept a great interest for many years. The new theory is based on the Dirac theory of relativistic quantum mechanics. The method of calculation had not been so well established at that time, so the calculation was by no means easy. He started working on this problem in cooperation with Oscar Klein from the spring of 1928, and was able to finish the work [5] in the summer. The Klein-Nishina formula thus obtained is really a gem of quantum mechanics to be

Fig.1. Nishina at Copenhagen, with his friends. 1925. From left to right : Nishina, Dennison (U.S.A.), Kuhn (Switzerland), Kronig (Holland), Ray (India).

remembered in history. I do not dwell on this topic any further, because Professor Ekspong will talk on this subject.

So Nishina finally left Copenhagen in October, 1928, cherishing the happy memory of the time spent at Copenhagen with Niels Bohr and many physicists who were fated to carry on the further revolution of modern science. He returned home in December of that year after making visits to several places in the United States.

Having returned to Japan, he joined Nagaoka's group at RIKEN. He must have felt like Urashima Taro or like Rip Van Winkle after so many years of absence from his mother country. There too, science had been progressing, but the atmosphere of the society was not yet as mature as in advanced countries. He had to be patient to realize his ambitions. It took a few more years until he was promoted to a chief researcher in RIKEN and in 1931 he started to build his own laboratory.

It was very fortunate that in 1929 he was able to invite Heisenberg and Dirac who accepted the invitation to come to Japan on their way back to Europe from America. Their visit gave great excitement to Japanese physicists, particularly to the younger generation. Nishina made a great effort to arrange their visit, to help the audience to understand their lectures. All of this made Nishina's presence more impressive among his Japanese colleagues. He was invited to the universities at Kyoto and Hokkaido to lecture on the new quantum mechanics. At Kyoto he met two young physics students, who regarded

Nishina as their teacher and later became to play the role of his successor and to lead modern physics in Japan. They were H. Yukawa and S. Tomonaga.

In 1931, Nishina started his laboratory. As the subjects of his research program he chose

(1) quantum mechanics,

(2) nuclear physics,

(3) studies of atoms and molecules by X-ray spectroscopy,

(4) use of spectroscopy for chemical analysis and its applications.

The list was revised the following year. The items (3) and (4) were replaced by

(5) study of cosmic rays,

(6) generation of high energy proton beams.

This list shows what Nishina intended to develop in Japan, in order to bring its science from the state of an undeveloped country to that of advanced societies like Europe and America.

It was fortunate for Nishina that he was able to build up his laboratory in a relatively short time. This was only possible because RIKEN at that time was a unique institution. It was very young, less than fifteen years old, and was an entirely new system independent of the government and existing universities. Originally it was planned to raise money for research from industry, which naturally turned out to be unsuccessful. Dr. Ookouchi, the third president of RIKEN was an eminent administrator. He created a number of companies to use the inventions made by RIKEN researchers. Many of these companies were very successful and brought a considerable amount of research money back into RIKEN laboratories. Dr. Ookouchi used to tell his researchers not to worry about money but only about their work. Therefore RIKEN had an extremely active atmosphere and it was called the paradise of researchers. It is said that Nishina's spending was always much more than his budget. Universities at that time were extremely poor regarding research money. Thus we see Nishina's projects progressed unusually fast with the strong support of RIKEN. Nishina was able to recruit brilliant young researchers so that his laboratory grew up very fast. The number of researchers at its maximum exceeded one hundred. A laboratory of this size was never possible in a university or in any other institute.

Let us now survey briefly how Nishina actually proceeded to achieve his objectives. Apparently he greatly emphasized theoretical physics. Fortunately, he was able to invite very able young researchers to his group. They were Shinichiro Tomonaga, Shoichi Sakata, Minoru Kobayashi, Hidehiko Tamaki and others. Yukawa was not in this group for he joined the new science faculty of Osaka University, but he kept up good contact with Nishina's group. Nishina himself was not able to do theoretical work as much as he probably wished, because his main efforts turned soon to experimental work. But in collaboration with Tomonaga and Sakata he studied pair creation probabilities by photons [6]. As Prof. Kobayashi recollects, Nishina guided the research work in much the same way as that in Copenhagen. Namely, discussions between researchers were regarded as most important. This style of work was new among the Japanese researchers at that time. Like Bohr, Nishina was able to encourage and train younger researchers. Tomonaga writes in one of his recollections, that he often became pessimistic about whether he was talented enough to do theoretical physics and it would be better to quit. But every time, Nishina warmly encouraged him to recover confidence in his ability. Yukawa also recalls Nishina being like his loving father. When Yukawa got the idea of the meson mediating the nuclear forces [7], Nishina was one of the few who immediately recognized its importance and gave the strongest support. Although the theoretical group of Nishina's laboratory was not so big, it was the most active in Japan and was influential in developing the quantum theory in Japan.

Nishina regarded cosmic ray research as the key subject to start new experimental physics. With a few members of the laboratory he was able to improve the counters and cloud chambers and set up observatories at various places like the top of Mt. Fuji, Shimizu Tunnel and so forth. Great excitement occurred with the finding [7] of a track in 1938 a little later than that made by Anderson, and Neddermeyer which was supposed to be the evidence of a Yukawa particle but later was proved to be another kind of new particle now called a muon. Nishina's group conducted a considerable amount of work in cosmic ray physics and laid the foundation of a strong tradition of cosmic ray research in Japan [8].

However, the greatest effort of Nishina was related to starting up nuclear physics in Japan, and in particular to the construction of cyclotrons. From 1930, the frontier of physics had shifted to nuclear physics with the use of various kinds of accelerators. In 1935, RIKEN decided to start a nuclear physics program with the cooperation of Nishina and Nishikawa Laboratories. A small cyclotron with a 23 ton electromagnet was successfully constructed by the Nishina group in 1937 [9]. Using this, the researchers irradiated all kinds of elements by fast neutrons. This was important work in nuclear physics. Beside this, radio-biological studies were initiated in Nishina's group by a team of biologists in cooperation with physicists and chemists [10]. This belongs to the earliest work of radio-biology. Thus, Nishina is regarded as one of the pioneers in this field.

The small cyclotron was successful, but it did not satisfy Nishina's ambition. He wanted to build a large cyclotron, about ten times as large as the small one. Through Dr. Sagane, the son of Prof. Nagaoka, a young member of the Nishina laboratory, Nishina learned that Lawrence at Berkeley was considering a project similar to his idea. So cooperation began between Nishina and Lawrence with a deep friendship between two physicists who had never seen each other before. By this cooperation, Nishina was able to buy a big electromagnet from an American company, which was the same make as that used by Lawrence. Unfortunately, however, the construction thereafter met

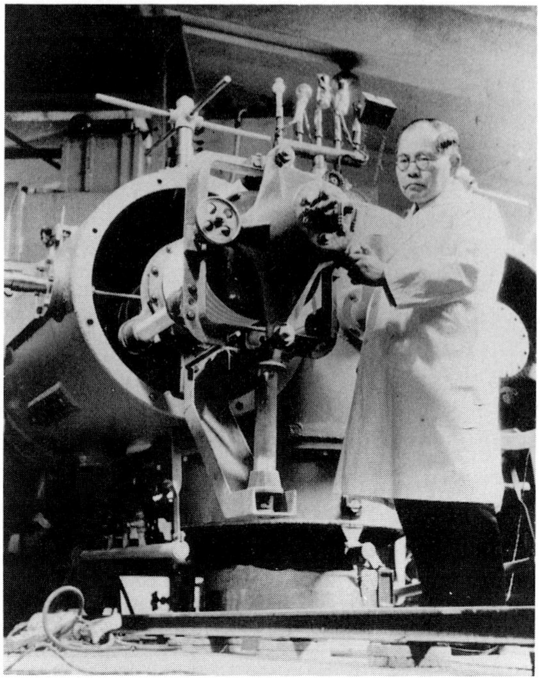

Fig.2. Nishina standing in front of the "large cyclotron", 1943.

great difficulties. It took much longer than Nishina expected. The difficulties were even bigger, since Japan was in the war by this time. Nishina did not give up the project. After great effort, the beam finally came out of the big cyclotron in February 1944. Although it was behind Lawrence by more than three years, this cyclotron was the second biggest in the world when it started to operate, of which Nishina had a right to be proud. When he left home for Europe in 1923, there was no nuclear physics at all in Japan, which was far behind in modern physics. But now he was happy to have put his country in second place, behind the United States. However, the story of the large cyclotron was a tragedy, to which I shall come back later.

In 1937, Niels Bohr visited Japan with his family in response to Nishina's long-standing invitation, and gave great encouragement to Japanese scientists. It was almost the last days of the happy time, because Japan was becoming internationally isolated by her militaristic policies, finally rushing into the reckless war. During the wartime, Nishina's group had to engage in the project to develop the nuclear bomb. The scale of the project was nothing compared to the Manhattan project and the researchers concluded that it was impossible to produce a nuclear bomb within fifty years. Japan was already losing the war then, and finally there came the disasters of Hiroshima and Nagasaki. Nishina was sent there to investigate if the bombs were really atomic. His report must have been a decisive factor in the political leaders' final decision to surrender.

Fig.3. Professor N. Bohr and Mrs. Bohr, visiting Japan, 1937, at a garden party at Takamine's home.

When the war came to an end, there was almost nothing left of Nishina's ten years' work. His laboratory at RIKEN was bombed and the small cyclotron was burnt. The big cyclotron fortunately escaped the damage. A few months after the surrender, the American occupation army came into Nishina's laboratory unexpectedly, broke the machine and sank it in the Tokyo Bay. Later the Secretary of War admitted that the destruction was a mistake by the War Department of the US Government. Even if it was an unfortunate accident, it still discouraged Nishina from resuming his scientific activities. However, this was just the beginning of the disaster.

The following year, in 1946, RIKEN itself was ordered by the Occupation Force to dissolve. The RIKEN family of companies supporting RIKEN, namely the RIKEN CONCERN was considered as something similar to zaibatu, undesirable for demilitarization of the Japanese economy. So RIKEN had to seek some way of living by itself. Since Nishina was the most distinguished among the remaining senior researchers, he had to take the full responsibility for reconstructing RIKEN from its fragments. Thus he had to give up science to become an administrator to earn money for the researchers, pay their salaries, and research money. In the postwar economy in complete social disorder, the task was difficult beyond our imagination. After a great struggle and great efforts, he managed to create a company with the name of RIKEN Co. He managed to construct a production line of penicillin within RIKEN, which was fairly successful at supporting financially the research activities at RIKEN. He became the president of the RIKEN company. Thus he saved RIKEN from collapse. If there were no Nishina, RIKEN would not have survived.

The great effort of Nishina to reconstruct RIKEN out of disaster was a part of his sacrifice to save his country . He felt very strongly his responsibility as the most influential leader in science. Building up of science and technology was the slogan of the Japanese to reconstruct the country from ashes. So he was obliged to extend his activities beyond RIKEN to the problems of the whole country. When the Science Council of Japan started in 1948, he was elected a member of the Council and then took up the responsibility of Vice-Presidency. In order to restore scientific international cooperation, he was sent by the Science Council to the General Assembly of ICSU which took place in Copenhagen in 1949. Reunion with Niels Bohr and his family was the greatest pleasure after such terrible years. The year of 1950 brought him happy news. Through the good will of American scientists, particularly Dr. Harry C. Kelly, who was a science adviser at the General Head Quarters of the Occupation Army doing his best to encourage Japanese scientists, the import of radioactive isotopes was made possible. Although the cyclotrons were lost by the war, Nishina now was able to start radio-isotope work, to encourage his fellow researchers to start working in many important fields of physics, chemistry, medicine and biology.

Unfortunately, his health was already deteriorating through overwork. It was January 10th of 1951 when he closed his 60 years' life of dedication to science and his beloved country.

Dr. Nishina died too early. In the forty years after his death, Japan has changed greatly. If not to our own satisfaction, modern science and technology have made enormous progress in Japan. Even though the number of Nobel laureates is still too small among the Japanese, the basic level of its modern science is ranked highly with the most advanced countries. And Japan's advanced technology in application to modern industries is very remarkable. What would Nishina say, if he was still alive and saw today's Japan ?

Nishina lived a life in a transient era of history that was most drastic and dramatic. It was the time when quantum theory cast off the older skin of classical physics, and atomic physics shifted to nuclear physics. Older concepts were revolutionized by new concepts. When Nishina started studying physics, Japan was only a developing country in the far east, far apart from the center of western civilization and the center of science revolution.

Although Japan's physics was steadily progressing before Nishina's homecoming, as is shown by some significant achievements by Japanese physicists in the 1920's, for instance, in X-ray crystallography, atomic spectroscopy, and electron diffraction experiments, the geographic distance and still backward technological level were great barriers hindering its ability to catch up with physics in the advanced West. Nishina was the right person with the destiny to bridge the gap between the older Japan before 1920 and the modern Japan after 1930. His role could be compared to that of Rabi and Oppenheimer in the United States. In fact, the growth of Japanese science and technology in the 1930's was remarkable. Industries were growing. National universities were created. Higher education was leveling up. Nishina was destined to lead modern science in Japan. As he might have foreseen when he chose physics after studying electrical engineering, he was successful in his pursuit of his objectives. If it had not rushed into the reckless war, Japan would have been able to attain a reasonably advanced level of modern sciences by the 1950's before Nishina's untimely death.

It is useless to talk about a historical if. But I only mentioned this to remember the great man with unusual talent who devoted his whole life to science and to his country.

Thank you very much for your attention.

References

[1] Special Issue for the Centennial Anniversary of Yoshio Nishina **Buturi** (Physical Society of Japan) October 1990.

[2] D.Coster, Y.Nishina and S.Werner : Z.Phys. **18** (1923) 207.

[3] Y.Nishina : Phil. Mag. **49** (1925) 521. Y.Nishina and B.Ray : Nature 117 (1926) 120.

S.Aoyama, K.Kimura and Y.Nishina : Z.Phys. 44 (1927) 810.

[4] Y.Nishina and I.I.Rabi : Verhandl.Deutsch.Phys. Gesell. 3. Reihe 9. Jahrg (1928) 6.

[5] O.Klein and Y.Nishina : Z.Phys. **52** (1929)853.

[6] Y.Nishina and S.Tomonaga : Proc. Phys-Math.Soc.Japan Third Series **15** (1933) 248.

Y.Nishina, S.Tomonaga and S.Sakata :Suppl.Sci.Pap. I.P.C.R. **17** (1934) 1.

[7] H.Yukawa : Proc.Phys.-Math Soc. Japan **17** (1935) 48.

[8] Y.Sekido, in Early History of Cosmic Ray Studies, ed. Y.Sekido and H.Elliot (D.Reidel Dordrecht 1985) p.187.

[9] Y.Nishina, T.Yasaki, H.Ezoe, K.Kimura, and M.Ikawa : Nature **146** (1940) 24.

Y.Nishina , T.Yasaki, K.Kimura and M.Ikawa : Phys.Rev. **58** (1940) 660 ; ibid **59** (1941) 323, 677 ; Z.Phys. **119** (1942) 195.

[10] Y.Nishina and H.Nakayama : Sci.Pap. I.P.C.R. **34** (1938) 1635.

Y.Nishina, Y.Sinoto and D.Sato : Cytologia, **10** (1940) 406, 458 ; ibid **11** (1941) 311.

Yoshio Nishina, a Founder of Modern Science in Japan

J. Kondo

Science Council of Japan, 22-34, Roppongi 7-chome,
Minato-ku, Tokyo 106, Japan

1. Introduction

Yoshio Nishina was born in Okayama prefecture, Japan, as a son of a rich family on 6th December, 1890. He found employment in the Institute of Physical and Chemical Research (RIKEN) after graduating from the Department of Electrical Engineering, the Imperial University of Tokyo in 1912. He stayed about a year at the Cavendish Laboratory in Cambridge under E.Rutherford from 1921, then moved to the University of Göttingen and finally he joined N.Bohr's school and stayed in Copenhagen from 1923 to 1928.

He returned home in 1928 and distinguished scientists in the field of theoretical physics, atomic nucleus and cosmic rays studied under his leadership at RIKEN. He became the director of RIKEN in 1948.

He was well known as a leading scientist among the physicists of Japan. In 1949, he was elected as a member of the first term (1949-1951) of the Science Council of Japan (JSC) and stayed as Vice President of JSC until he passed away on 10th January in 1951 [11].

2. Nishina and Bohr

Nishina studied at Bohr's Institute at Copenhagen for 5 years, in the period of the birth of modern physics. There is no doubt that he was very much influenced by Bohr, not only in the study of nuclear physics but also by the atmosphere dominating Bohr's school. He learned the importance of freedom of thought, which is a basic factor for creativity.

In the spring of 1937, Niels Bohr came to Japan and spent about a month there. He gave a lecture under the title "On the uncertainty principle and the structure of atomic nucleus" at the auditorium of the Kyoto Imperial University (KIU) on 10th May. I had a chance to attend this lecture when I was a student at the Department of Mathematics at KIU [1] [10].

Nishina stood on the right side of the tall Bohr as an interpreter. Bohr spoke gently, occasionally writing a short formula on the board. Nishina trans-

*President of the Science Council of Japan

Springer Proceedings in Physics, Volume 57
Evolutionary Trends in the Physical Sciences Eds.: M. Suzuki and R. Kubo

lated to hundreds in the audience, who were listening calmly to the lecture by the famous Nobel laureate. Suddenly Nishina stopped the interpretation and started a discussion with Bohr on the formula on the board. I could not understand what was wrong. After a few minutes of intermission, Nishina restarted the interpretation. I realized that Nishina was not merely translating but he was giving the lecture in his own words. It was really a fascinating sight showing vividly the atmosphere of Bohr's school. As a matter of fact, however, the lecture was too difficult for a first-year student at the University. I cannot remember the details of the lecture at all.

3 Nishina and the Atomic Bomb

In the morning of 6th August 1945, an atomic bomb was detonated above the city of Hiroshima and almost all parts of the city were completely destroyed instantaneously.

The military headquarters asked Nishina to inspect the city in order to identify the bomb of ultra strong destructive power. Nishina arrived in Hiroshima on 8th August. He immediately judged that it was an atomic explosion with strong radiation when the sensitized X ray films were shown to him at the Hiroshima Red Cross Hospital.

After his arrival in Hiroshima, he visited several places and walked around the ruin. He encountered a miserable scene of thrill and horror which he had never seen before. His body was exposed to strong radiation without any protection for a couple of days while he stayed in Hiroshima. It is estimated that this exposure might have caused the cancer which led to his death due to liver cancer five years later.

The army asked Nishina to study the possibility of developing an atomic bomb in Japan. He concluded in 1943 that in principle it might be possible, but in practice was not possible, because of the shortage of materials and skilled labor. It was difficult to find a vacuum pump powerful enough for condensing uranium, according to his experience in constructing cyclotrons. Still the infrastructure of industry was insufficient to challenge the frontier of technology. The gap between science and industry was large in many cases, and innovative technology was not mature in the war period. This situation has improved considerably in recent years.

After inspecting the miserable states in Hiroshima and Nagasaki, Nishina considered very seriously the responsibility of scientists. He recognized that science may contribute to the development of weapons, and consequently that scientists should make every effort to avoid a war [7].

4. Nishina as a Member of JSC

The Science Council of Japan started its activities from the First General Assembly in January 1949. The special law for JSC had been passed by the Diet in the previous year. JSC is a governmental organization representing all qualified scientists in Japan. One of the objectives is to rebuild the country on the basis of science. It aims to utilize the fruits of science to improve the living standards of the people, the administration of enterprises and the political decisions of the government. The period of service for one term is 3 years. The members, totalling 210, were elected by the direct vote of the qualified scientists until 1985 when the process of nomination was completely changed. Presently the members of JSC are nominated through the scientific professional societies [2].

Science, or Gakujutsu in Japanese, has a broader meaning than the English word science. It represents all the intellectual activities of human beings. Therefore, JSC includes not only pure and applied science but also cultural and social sciences. JSC consisted of seven divisions, namely, divisions of cultural science, jurisprudence, economics, natural science, engineering, agriculture and medical science including pharmacology. About 30 members are allocated to each division [6].

An organization of this kind is useful for approaching problems in an interdisciplinary way, but takes a long time to come to conclusion due to lack of knowledge in the background of the colleague members.

Nishina played an important role in determining the basic rules of management of JSC. The General Headquarters (GHQ) of the occupational forces recommended to the government of Japan to reform almost every system including administration, education, etc. in order to encourage democracy everywhere.

The recommendations by GHQ had a superpower in the period of occupation but some officers of the GHQ were very kind and capable of finding an ideal system for cooperation with Japanese counterparts. Dr. Harry C.Kelly arrived in Japan in January 1946 as a scientific adviser to GHQ. Dr. Kelly respected Nishina for his enthusiasm in scientific studies and then became a good friend to him [13].

Japanese scientists were trying to reform the administrative system on science. Dr. Kelly helped substantially in the reformation of the system. The Imperial Academy, the Scientific Research Council, and the Society for Promotion of Science were restructured after the JSC was organized [4] [5] [8] [12].

5. Nishina and Riken

After coming back from Europe via the U.S.A., Nishina returned to RIKEN and started to organize his research group as a center of modern physics in

1928. It was the time when quantum dynamics was established. In imperial universities, few posts are available to include a new field of science. Nishina preferred to stay in the private sector where he believed that academic freedom could be sustained.

Nishina organized a research group with R.Sagane and M.Takeuchi on experiments and S.Tomonaga and S.Sakata on theory in 1931. This group increased gradually with the development of nuclear physics. Nishina constructed a Wilson's cloud chamber of large scale and observed a trace of a large particle in a cosmic ray.

In 1938, Nishina built a cyclotron with the help of E.O. Lawrence, University of California. Immediately after the completion of the cyclotron, Nishina took up another project of constructing a large cyclotron of 850 MeV. This was a really difficult task. Large scale magnets and vacuum pumps with high performance are needed to construct a cyclotron of large scale. The level of Japanese mechanical industries was not high enough to manufacture such machines of high quality.

He concentrated his effort on this project. In the meantime scientists in his group insisted on conducting experiments using the smaller cyclotron. They complained that their boss was always too ambitious for challenging difficult tasks.

Nishina worked hard and indicated an excellent leadership for the completion of the larger cyclotron. It was just before the start of the World War, Nishina and his scientists could not continue experimental work due to a shortage of resources and manpower.

In 1945, when the allied forces occupied Japan, GHQ destroyed these machines in order to abolish the research on atomic weapons in Japan. It would be not difficult to imagine how disappointed Nishina was when he was informed that GHQ had destroyed his cyclotrons owing to a misunderstanding [9] [14].

Acknowledgments

The present author never studied under the guidance of Nishina. He attended only once the open lecture by Bohr when Nishina appeared as an interpreter. He is indebted greatly for the story of the inspection tour in Hiroshima to Mr. Kikuo Nishida who graduated from the Department of Physics, the Imperial University of Tokyo and worked together with Nishina as a technical lieutenant at Kure Base of the navy. He also learned a lot about Nishina through Professor Seitaro Nakamura who studied theoretical physics under Nishina at Riken [3] [15].

References

[1] The Kyoto Imperial University News, May 5th 1937.

[2] Reorganization of Science and Technology in Japan. Report of the Scientific Advisory Group to the National Academy of Science, United States of America 1947

[3] Nishina, Y. News Paper The Yomiuri, 2nd June, 1948

[4] Proceedings of the General Assembly, JSC 1949

[5] Recommendations and Appeals Vol. 1, JSC 1949

[6] Yuasa, M. : Science 50 Years (The Jiji Tsushin 1950)

[7] Tomonaga, S. and Tamaki, H. ed. : Nishina Yoshio (Misuzu Book Co. 1952)

[8] 25 Years of the Science Council of Japan, JSC 1974

[9] Tamaki, H. : Nishina Yoshio, Biographies of the World, Vol. 4 (Kokudo Co. 1976)

[10] The 70 Years History of the Kyoto University (Kyoto University 1980)

[11] Nakamura, S. : Nishina Yoshio, Hisashi Inoue and Others : Distinguished Japanese in the Last 100 Years, Vol. 1. (The Yomiuri Press Co. Ltd. 1985)

[12] Fukushima, Y. : 40 Years of Scientists' Forest - with the Science Council of Japan (Nippon Hyoron Sha 1986)

[13] Yoshikawa, H. : Science Has No Boundary of Nations, Memorabilia of Dr. Kelley (Mita Pub. 1987)

[14] Imamura, M. : Cyclotrons at the Institute of Physical and Chemical Research. - A Cyclotron Can Be Destroyed, But Not Defeated. - J. Kroh ed. "Early Developments in Radiation Chemistry" (Royal Society of Chemistry 1989) pp. 245-255.

[15] The Centenary of the Birth of Yoshio Nishina Special Issue BUTSURI, Vol 45, Oct. 1990.

In Memory of Dr. Yoshio Nishina

M. Kotani

Japan Academy, Ueno Park, Taito-ku, Tokyo 110, Japan

It is a special honor for me to say a few words on this anniversary, Nishina's centennial. My task here is to talk mainly about him around 1930, when I saw him fairly often, but I am afraid my memory of more than sixty years ago is rather hazy. Fortunately Dr. H. Tamaki has provided me with many documentary materials kept in RIKEN to aid my memory, and I am greatly indebted to him.

Now, with your permission I would like to explain briefly how we studied the new quantum mechanics at the Tokyo Imperial University in the late 1920s. I was an undergraduate student in physics for three years from April 1926 till March 1929. For our first two years there were no lectures on quantum mechanics. But since this new physics seemed very attractive to us, a voluntary informal study group was formed in our class. Inui and I were most eager members; we had been classmates since high school days. Because no textbook of quantum mechanics was available, we had to read some original papers in "Annalen der Physik," "Zeitschrift für Physik,","Proceedings of the Royal Society," as far as they were understandable for us, and to explain the results at meetings of the study group for discussion. Through this voluntary study, we were familiar with elementary quantum mechanics before attending Professor Sakai's lecture in 1928.

In the course of such voluntary studies in 1927, Inui and I found an interesting paper by Heitler and London in an issue of "Zeitschrift für Physik." This paper explained quantitatively why two hydrogen atoms can combine to form a stable molecule, using antisymmetry of two electron spins and the Pauli principle. I was much impressed by this paper, realizing that the covalent bond in chemistry, which had been a mystery for us physicists, was suddenly clarified on the quantum-mechanical basis. I thought physics and chemistry were united in physico-chemical or physical science. Later chemical physics and quantum chemistry developed on this basis. I believe Nishina welcomed this growth of quantum mechanics into the field of chemistry, although I do not find any direct statement by him on this matter.

Now, I have to talk about Nishina, who returned to Japan in December 1928 after a seven-year stay in Europe. The period when Nishina studied under N.Bohr coincided with the years of a revolutionary change in physics due to the birth and development of quantum mechanics. From his unique, wonderful experience, it was quite natural that Nishina eagerly wished to introduce

Springer Proceedings in Physics, Volume 57
Evolutionary Trends in the Physical Sciences Eds.: M. Suzuki and R. Kubo

a European atmosphere of devoted pursuit of truth to Japan to encourage young scientists in active research. For this purpose Nishina thought it highly desirable to invite Heisenberg and Dirac to Japan, and before leaving Europe, Nishina could confirm that they were willing to visit Japan if invited for a convenient period.

Nishina returned to RIKEN, Tokyo in December 1928. Fortunately it became clear in April that early September in 1929 would be convenient for both guests, and they were formally invited by Keimeikai– a foundation supporting public activities in sciences and humanities. This smooth progress of Nishina's plan was due to Professor Nagaoka's kind support of this plan and his effort to convince Keimeikai of its importance. The meeting consisting of lectures by Heisenberg and Dirac was assigned to be held on September 2-5, and was publicized by newspapers well in advance. RIKEN informed many universities in Japan of these lectures, hoping to arouse interest in new quantum physics among young scientists in various places. I know that during this summer voluntary study groups, as mentioned above, were organized in some universities to prepare for the lecture meeting.

On August 30th Heisenberg and Dirac arrived at Yokohama from San Francisco on the same boat. Participants were mostly from the Tokyo area, but a fair number of scientists came from other regions of Japan. For instance, Tomonaga traveled from Kyoto to attend, and I was introduced to him by Nishina.

The lecture meeting was held on September 2-5, and each lecturer, Heisenberg and Dirac, gave one lecture each day, i.e. four lectures as a whole. Some of the lectures were given at the university and the others in RIKEN. Heisenberg's four lectures included "Theory of Ferromagnetism (H1)," "Theory of Electric Conduction (H2)" while Dirac gave lectures on "Statistical Quantum Mechanics (D1)", "Quantum Mechanics of Many Electron Systems (D2)" and two others. I was particularly interested in two lectures (H1) and (D2), which were closely related through electron spins and the Pauli principle, and may be regarded as extensions of Heitler-London's theory of the covalent bond in the hydrogen molecule.

Throughout this meeting Nishina attended all the sessions, and orally translated the lectures into the Japanese language, paragraph by paragraph. This must have been a very difficult task, but contributed very much to a good understanding of the rather difficult content of the lectures.

After this exciting meeting the guest lectures made a short trip to Kyoto and Nara, and left Japan for Europe, Heisenberg via the Suez Canal and Dirac by the Siberian railway. After they left Japan Nishina wanted to provide the text of their lectures in the Japanese language for wider circles of Japanese students and scientists, and engaged himself in translation, taking time off from his busy research and related business work. In the last stage, Inui and I were asked to assist him in finalizing the manuscript and proofreading. This Japanese volume was published in 1932 as the eleventh bulletin of Keimeikai, and about 150 copies were distributed on request.

What I have talked today about Nishina is a very small portion of all his activities, but it is clear that Nishina did the best in planning and executing the lecture meeting that he could. We can recognize his devoted love of science and of Japan, even from what he did for the lecture meeting although he had no laboratory of his own in RIKEN yet. As a by-product of his endeavor he became familiar to many physicists in Japan, and he was invited to Hokkaido and Kyoto universities to deliver lectures in 1931. Later in the same year, he was appointed a member of RIKEN, having his own laboratory, and it became possible to invite young able physicists to his laboratory to carry out research as he wanted. But I would like to close my talk here. Let us try not to lose memories of our great physicist Nishina, devoted to the progress of physics in Japan, sometimes under very difficult social conditions.

To the Conference Commemorating the Centenary of Yoshio Nishina

A. Bohr (Read by B.R. Mottelson)

Niels Bohr Institute, University of Copenhagen, DK-2100 Copenhagen, Denmark

The Niels Bohr Institute has asked me to send greetings to the Nishina Centenary Meeting. This is a welcome occasion for us to commemorate Nishina's long association with our Institute and the warm personal relationship that developed between Nishina and my father. These associations and their continuations fall into three periods, beginning with Nishina's participation in the research at what was then the Institute for Theoretical Physics in Copenhagen, followed by Nishina's vigorous efforts to develop modern atomic science in Japan, during which time he persuaded my father to come for a visit. Later, after Nishina's death, the traditions of close cooperation by Nishina's successors and our institute has been a profound source of inspiration to so many of us in the following generations.

Nishina, who had come to Europe for studies in 1921, met my father in Cambridge the following year and, in March 1923, wrote to him expressing the wish to study atomic physics in Copenhagen. His chief wish, as he says in the letter, was "to study your theory of spectra and atomic constitution in detail. But if anyone wants assistance in the experiment or the calculation I should do it with pleasure". As it turned out, he took a leading part in the experiments at the Institute concerned with various aspects of X-ray spectroscopy, which at that time was contributing so much to atomic theory and opening up new connections to chemistry. At that same time, Nishina immersed himself in the revolutionary theoretical developments that were taking place in those years in quantum physics. Thus, with the advent of Dirac's theory of the electron, he could immediately, together with Oskar Klein, work out the fundamental cross-sections for the scattering of electromagnetic radiation by electron. The correspondence from those years between my father and Nishina shows how much Nishina's participation in the work of the Institute was appreciated. This applies not only to his own work, but also to the help that he was able to give to his colleagues, on the basis of his broad interests and experience combined with his kind and generous personality.

Nishina left Europe in October 1928 and, on his return to Japan, was able with remarkable effectiveness to promote the development of atomic science in Japan. In this connection, he almost immediately began to make arrangements for a visit by my father. The visit was originally planned for 1930, but had to be postponed several times until it took place in the spring of 1937, thoughtfully arranged to coincide with the season of the cherry blossoms. My father was

Springer Proceedings in Physics, Volume 57
Evolutionary Trends in the Physical Sciences Eds.: M. Suzuki and R. Kubo

joined by my mother and my older brother Hans. From the vivid accounts they gave on their return, supplemented by a film taken by my father, which he enjoyed showing with lively comments, I feel almost as if I had been present. Nishina, helped by other former collaborators of the Institute in Copenhagen, had made great efforts to arrange a program that would enable my father to have broad contact with scientific circles in Japan and at the same time to experience the uniqueness of Japanese culture in its great variety. This meeting had deep consequences as testified by the reactions from both sides. For my father, his first contact with the orient was a great inspiration, which became an important element in his views of the human situation. Throughout his later years, when expounding his views in discussions and conversations, he would often refer to humor and wisdom from the East and would color his arguments with stories from Japan. He felt a certain kinship between his own way of thinking and elements in the oriental philosophy of life. Thus, a favorite quote was from "ancient thinkers in the East" reminding us that "in the search for harmony in human life, we should never forget that we are ourselves actors as well as spectators" in the great drama of existence. In those years, my father was occupied by the question whether human qualities manifesting themselves in different cultures are inherited, and he advocated the views that a human culture is to be compared with a flower that can grow in a variety of different environments. He brought up these issues in Japan, where they gave rise to lively discussions. Through the personal contact with his Japanese friends, despite the enormous differences in cultural tradition, he found a confirmation of his views.

After the war, Nishina devoted all his efforts to rebuilding the institutions and facilities for scientific research in Japan. His untimely death in 1951 was felt as an irreparable loss not only to the Japanese science community but also to his colleagues and friends abroad, and I remember how strongly my parents felt the personal loss they had suffered. But the spirit of Nishina continued to be an inspiration for those following after him, who took up the task of promoting science in Japan and strengthening connections to the international community. The appreciation of Nishina's role in the development of science in Japan found expression in the establishment of the Nishina Memorial Foundation. This Foundation played a major part in re-establishing the close ties between our Institute in Copenhagen and the physics community in Japan. At a later stage, the Nishina Memorial Foundation, through its president Professor Sin-itiro Tomonaga, took the initiative to obtain major support from the Commemorative Association for the Japan World Exposition to the Niels Bohr Institute to enable the Institute to maintain its function as a meeting place for scientists from all over the world. In the life of the Institute in the last decades, the participation of the new generations of Japanese physicists, both experimental and theoretical, has been an increasingly important element. Thus, in our daily life, we feel the continued inspiration and importance of the ties that were established when Nishina was a member of the Institute in Copenhagen more than 60 years ago.

Oskar Klein and Yoshio Nishina

G. Ekspong

**Department of Physics, University of Stockholm,
Vanadisvägen 9, S-11346 Stockholm, Sweden**

Abstract. The joint work in 1928 between Oskar Klein and Yoshio Nishina at Niels Bohr's Institute in Copenhagen is described. The scientific background to Compton scattering is briefly reviewed and the derivation of the famous Klein-Nishina formula discussed as well as the subsequent development of the subject.

1. Introduction

Yoshio Nishina arrived in Copenhagen in 1923 at the age of 32 years. Oskar Klein, who by then was 28 years old, had been with Niels Bohr for 5 years, but left in 1923 to return for good only in March 1926. Nishina stayed in Copenhagen with some short interruptions until October 1928, when the two papers on the Compton effect were ready for publication. Klein on the other hand remained with Niels Bohr until 1930, in which year he was appointed full professor at the University of Stockholm, at that time a private university.

The research activities of Klein and Nishina were different until 1928. Yoshio Nishina had done experiments on Compton scattering before arriving in Copenhagen during his visit to Rutherford's laboratory at Cambridge in England. At Copenhagen he did experimental X-ray spectroscopy for most of the time. Nishina switched in 1928 to theory for his collaboration with Klein. Since the work by Yoshio Nishina is being reviewed at some length during this symposium, I will recall some aspects of Oskar Klein's early career before discussing the Klein-Nishina collaboration.

Let me also shortly mention my own contacts with Klein. We overlapped as faculty members at Stockholm University for about two years until his retirement in 1962. However, his presence was enjoyed for many years after his retirement. I had done experimental studies of antiproton-nucleon annihilations since 1956, i.e. for 4 years, when I became faculty colleague with Klein. In those years his focus of interest was cosmology. About this time he formulated his ideas of a matter-antimatter symmetric metagalaxy. We held joint seminars on various aspects of this problem with the participation of theoretical and experimental physicists and astronomers. Klein did not like big words, such as a theory for the "Universe", since according to him our known part of it, the metagalaxy, may not be the whole world. His attitude was always somewhat cautious, he seemed to regard theories as ten-

Springer Proceedings in Physics, Volume 57
Evolutionary Trends in the Physical Sciences Eds.: M. Suzuki and R. Kubo

tative and never definitive. He was, like Niels Bohr, a philosopher of science, partly forced to it when trying to explain the nature of modern physics to the misunderstandings prevailing among some Swedish professional philosophers. Oskar Klein was a great man, honest and kind, humble and soft spoken.

2. Some early work by Klein

Apart from the Klein-Nishina formula, Oskar Klein is known for several other achievements, among them the Kaluza-Klein five-dimensional theory, the relativistic Klein-Gordon equation, the Jordan-Klein second quantization and the Klein paradox. Klein did, however, no joint work with either Gordon or Kaluza. In these cases there were independent papers with similar content. With Nishina the collaboration was very close and they signed the first paper jointly. The second paper, devoted to a treatment of polarization phenomena was signed by Nishina alone.

Klein was admitted at the age of only 16 years to the Nobel institute for physical chemistry in Stockholm two years before entering the University [1]. The head of this institute was Svante Arrhenius, famous for his dissociation theory of electrolytes, who had been awarded the Nobel prize for chemistry in 1903. Klein's first four papers are of a chemical nature. His Ph.D. thesis at Stockholm University in 1921 was again in chemistry and dealt with the statistical theory for suspensions and solutions, containing a generalised treatment of Brownian motion.

When arriving in Copenhagen Klein had already studied a great deal of quantum theory on his own. His first physics paper dealt with X-ray scattering from a plate. With S. Rosseland he predicted the existence of atomic collisions of the so-called second kind. As early as the early 1920's, Klein began a search for a deeper foundation for the quantum rules. He entertained the thought that wave interference could lie behind the appearence of integer quantum numbers. He found wave-fronts associated with classical particle trajectories in Hamilton's original approach to what became the Hamilton-Jacobi equation. Klein set out to derive the general Hamilton-Jacobi equation for a charged particle moving in a combined gravitational and electromagnetic field and saw a similarity with a wave in a four-dimensional space. Klein hesitated to complete his work, since he was disturbed by the possibility that non-linear terms would be present. Klein's ambition was perhaps too high, trying to unite gravity and electromagnetism in a five-dimensional space-time and with the same stroke introduce a wave-theory for particles. The beginning of all this was done before Schrödinger's paper for the electron in the hydrogen atom appeared. Klein later completed his five-dimensional paper, and suggested that space in the fourth direction is closed to a very small circle, which he connected with the fact that electric charge is quantized [2, 3]. Pauli told Klein that a five-dimensional theory had been proposed by Kaluza [4] some years earlier.

During a short stay in Leiden in 1926, Klein started together with Uhlenbeck to calculate the Compton effect in the Schrödinger way, having seen Dirac's paper on the subject, done in the Heisenberg way. This calculation was, however, never finished. Klein instead began work on a correspondence treatment of wave mechanics. In that paper [5] Klein applied his relativistic wave equation and stated the wave-mechanical expressions for the densities of charge and current. He also included a treatment of the Compton effect. However, he did not calculate the scattering cross-section, but limited himself to showing that wave-theory leads to the same kinematical relations as those set up by Compton on the basis of energy-momentum conservation between colliding particles.

After Dirac's paper on quantization of the radiation field had appeared, Klein generalised it to quantize Schrödinger's matter field, which he published jointly with Jordan.

The first draft of Dirac's paper on the relativistic electron came to Bohr as an amazing surprise. Early in 1928 Bohr sent Klein to Cambridge to learn more about it. When Nishina in March 1928 came back to Copenhagen from a several months' visit to Hamburg his joint work with Klein began.

3. Compton scattering

It is hardly necessary to point out what a pivotal role Compton scattering has played in the history of physics. Einstein's vision about the existence of light quanta from his studies of energy fluctuations in thermal radiation in 1904 and his proposal in 1905 that light quanta would serve well to explain the photoelectric effect were resisted by many leading physicists for two decades. The wave-theory of light was not to be given up easily [6]. Compton's discovery of the frequency shift of scattered X-rays was also difficult to understand on the classical wave-theory. Many futile attempts were made to account for it. Compton's explanation (and the similar one by Debye) for the shift as due to energy-momentum conservation in each single scattering event between particles, one photon and one electron, was very successful but not convincing to everybody. The necessity to accept the photon concept became clear only after time coincidences and angular correlations between scattered quanta and electrons had been observed. The early confusion about how to remedy the inability of classical wave-theory to account for the observed angular distribution is fully exposed in the 1926 monograph by Arthur H. Compton [7]. Many attempts had been made in order to derive new scattering formulæ. Compton's first edition of the book appeared too early to report on the quantum theory treatments by Dirac and by Gordon, which were published in 1926, respectively 1927 [8,9]. Their formula agreed reasonably well with data, but was to be superseded by the one derived by Klein and Nishina.

4. The Klein-Nishina formula

The full paper by Klein and Nishina was published in German [10]. Only a short note with the result and a comparison with experiments appeared in English [11].

The starting point was Dirac's new relativistic theory for the electron. They referred to the earlier treatments by Dirac and by Gordon as being based on the 'older' forms of relativistic quantum mechanics. Klein and Nishina, however, did not use Dirac's theory of quantized electromagnetic fields. They choose to treat the field in a semi-classical way. They claimed that Dirac's radiation theory is expected to yield the same result as theirs, when one limits oneself to the first approximation. This was shown to be true independently by Ivar Waller [12] and by Igor Tamm [13] fully one year later.

Klein and Nishina discussed the problem of negative energy solutions to the Dirac equation and considered them physically without meaning. 'We will of course limit ourselves to positive values' was the content of one sentence in the paper. The field of the incoming electromagnetic wave was considered to perturb the electron. The charge current density was set up and treated so that the radiation field from it could be calculated. On this point they referred to the treatments by Klein and also by Gordon the year before. They arrived 'after some calculation' at a lengthy expression for the magnetic field at a point far from the electron. They pointed out that their result showed that spin-flip occurs besides no spin-flip.

They used the calculational rules given by Dirac for handling the matrices, in their case $\sigma_1, \sigma_2, \sigma_3$ as well as ρ_1, ρ_3 appearing in the Dirac equation and ρ_2 in the Lorentz transformations. When the magnetic field strength was squared they arrived 'after some calculations' at the expression:

$$\overline{H_0^2} = \frac{e^4}{m^2 c^4 r^2} \left(\frac{\nu'}{\nu}\right)^3 \left\{ \left(\frac{\nu}{\nu'} + \frac{\nu'}{\nu}\right) \epsilon^2 - 2(\mathbf{n}'\epsilon)^2 \right\}$$

where r is the distance from the electron to the point of observation and ϵ is representing the linear polarization of the incoming wave. The other factors are easily recognized from similar and more familiar expressions. The authors commented on polarization phenomena by saying that the scattered intensity does not depend on the initial polarization of the electron in this case of linearly polarized light. However, they pointed out that such a dependence appears in the case of elliptically polarized light, the treatment of which they referred to the subsequent paper by Nishina alone. In the joint paper they summed over the final state polarizations. They also introduced instead of ν' the scattering angle, which in their wave-mechanical language is the angle between the direction of observation and the normal to the wave front of the incoming light. The expression for the intensity as a function of this angle and also the angle between the electric vector of the incoming light and the direction of observation was found to contain one factor more than the corresponding expression given two years earlier by Dirac [8] and by

Gordon [9], a factor which becomes unity if the square of the energy ratio, i.e. $(h\nu/mc^2)^2$, can be neglected. They checked that in the classical limit of long wave-lengths their formula reduces to the one given by J.J. Thomson.

So far photons had not appeared in their paper, but were introduced towards the very end. In order to get the number of quanta scattered into the solid angle element, $d\Omega$, they multiplied their expression for the intensity by the factor $d\Omega/h\nu'$. Finally, an integration over angles led to the total cross-section ("scattering coefficient") as a function of photon energy.

In a short note [14] written by Klein in 1975, reminiscing his friendship with Yoshio Nishina, Klein tells how their joint work began. Gordon on a short visit to Bohr's institute suggested that the Compton effect might be a good problem for Nishina. Klein added 'and I, who had intended to attack that problem myself, agreed immediately'. Nishina started to read Dirac's paper and did that very carefully, so that when Dirac himself came to Copenhagen the same spring, Nishina told Dirac that he had found an error of sign in the paper. To this came Dirac's reply 'but the result is correct, however', followed by Nishina's reaction 'there must be two mistakes' and Dirac's final remark 'must be an even number of mistakes'.

Klein also recalls that there was not yet any established method to solve the problem, which made it difficult. At the beginning of the summer vacation period Nishina 'had — quite obviously — not been able to do more than a general preparation of the problem.' Klein then suggested a collaboration during the summer, when Klein and his family stayed at Lundeborg on the east coast of the Danish island Fyn, while Nishina was put up in a nearby pension. There is a story [15], told by Klein, that the two of them worked day by day independently during the summer, each sitting far from each other in folding chairs doing the lengthy algebra. After comparing notes at the day's end, they embarked on further calculations the following day. In early July in a letter to Bohr [16], Klein informed him that he and Nishina had begun to write a first draft during their joint vacation with the hope to have it finished quite soon. Although their joint short letter to Nature was signed on the 3rd of August 1928, Nishina wrote five days later in a letter to Bohr [17], that the work with Klein had not yet been finished. The plan was to resume the work later in August, when both of them would be back in Copenhagen, and finish it in September. The 16 pages long paper was received by Zeitschrift für Physik on the 30th of October.

In the subsequent paper [18], printed immediately after the first one, Yoshio Nishina derived formulæ for the polarization phenomena. It was submitted the same day as the first paper. Nishina, who had already been in France for several weeks, left Europe the following day. In his paper Nishina pointed out that the results were quite different from any earlier ones. At first he showed that the scattering of linearly polarized light on polarized electrons led to two incoherent, elliptically polarized scattered rays. This had to do with no spin-flip, respectively spin-flip transitions as he pointed out. The result was, however, independent of the initial spin state of the electron, as was mentioned already in the joint paper. He also treated the case of scattering twice at right angles, which was the usual geometry in experimental investigations of light polarization in Compton scattering. Therefore, he also calculated the scattering of an incoming elliptically polarized wave. The result was in this case clearly dependent on the initial spin state of the electron. Although Nishina immediately averaged over the initial spin directions, one can see here for the first time, that Compton scattering against polarized electrons is sensitive to the circular polarization of γ-rays.

A complete theoretical treatment of the general case was given in 1938 by W. Franz [19], who derived expressions for the scattering of elliptically polarized light against electrons polarized in an arbitrary direction. Following Gunst and Page [20] one often writes the differential cross-section for Compton scattering on polarized electrons as a sum of two terms

$$\frac{d\sigma}{d\Omega} = (\frac{d\sigma_{K-N}}{d\Omega}) + P(\frac{d\sigma_1}{d\Omega})$$

where the first term is the usual Klein-Nishina expression for unpolarized or linearly polarized light. The second term, which may add or subtract, depends on the circular polarization P of the photons.

5. The negative energy states

Klein and Nishina avoided explicit reference to electron states with negative energy in their derivation. Ivar Waller in Uppsala (now 92 years old) was the first to apply Dirac's radiation theory to the Compton scattering problem [12]. Somewhat later Igor Tamm in Moscow worked through the same problem [13]. Both of them used the original Dirac theory with empty negative energy states. The importance of not neglecting these among the possible intermediate states was emphasized. Waller did the sums over positive energy states and negative energy states separately and claimed that both are needed in order to obtain the Klein-Nishina formula. Furthermore, in the classical limit the negative energy states were found to dominate. Tamm was even more expressive in saying that one would get a grossly wrong result if the negative energy states were left out. He claimed that in the classical limit, scattering via positive energy states gives zero cross section, and that the finite Thomson cross-section derives via the negative energy states alone.

Waller knew about Dirac's idea to fill the negative energy sea with unobservable electrons and Dirac knew about Waller's results. In the new theory there would be a replacement of transitions to now unavailable negative energy states with other transitions in which a hole state (a virtual positron, in modern terminology) appears. In a letter to Bohr [21], Dirac wrote about this consequence of his hole theory in the following words: 'On my new theory ... there is ... a new kind of double transition taking place in which first one of the negative-energy electrons jumps to the proper final state with emission (or absorption) of a photon, and secondly the original positive-energy electron jumps down and fills up the hole with absorption (or emission) of a photon. This new kind of process just makes up for those excluded and restores the validity of the scattering formulas, derived on the assumption of the possibility of intermediate states of negative energy'. In view of this and the results by Waller and Tamm, one reaches the not so obvious conclusion that the low frequency, classical limit is dominated by intermediate states with three charged particles present, one of which is a virtual positron, the other two being the incoming and the outgoing electron [23]. In a reference to Dirac's new hole theory, Waller claimed, that as far as his paper was concerned, only formal changes would be needed. Shortly after, Dirac [24] showed this to be true.

The success of the Klein-Nishina formula thus led to a focusing on the importance of the curious, and problematic negative energy states, since they could not simply be ignored.

6. The Klein-Nishina formula as a tool for research

6.1 A tool for cosmic rays

In their paper Klein and Nishina suggested that their formula could serve as a tool to determine the energies of cosmic rays. One can write the total cross-section formula for Compton scattering in the extreme relativistic limit in the following form:

$$\sigma \approx \frac{\pi \alpha^2}{m E_\gamma} ln \frac{2E_\gamma}{m}$$

where E_γ is the γ-ray energy. It is thus obvious that if there were no other processes, high energy photons would be very penetrating and that the attenuation coefficient would be a measure of the energy. However, the scheme never worked. Another process turned up, pair creation, which had, of course, to do with the secret behind the negative energy solutions of the Dirac equation.

6.2 A tool which almost led to the discovery of the positron

Already when experiments to check the Klein-Nishina formula were being carried out it served as a tool. Reports came in 1930 from three places [25,26,27] that high energy γ-rays were attenuated much more than predicted by filters of high-Z materials, but not low-Z materials. Disagreements with the formula were taken as hints for the existence of some new process, the nature of which, however, was missed by those directly involved and by all others as well [28]. The new process was identified with the creation of electron-positron pairs only after the discovery of the positron by C.D. Anderson in 1932.

The first report of still another new phenomenon came in 1930 from Chung-Yao Chao, a Chinese visitor at the California Institute of Technology [29]. With hindsight his discovery of an isotropic radiation of about 0.5 MeV energy was identified with annihilation radiation. Chao returned to China shortly afterwards and now lives in Beijing, 88 years old. I had the pleasure to talk with him at length about his experiments during a visit to China in 1985.

In this instance the Klein-Nishina formula as a tool was on the brink of leading to the discovery of the positron.

6.3 Polarization of annihilation radiation

In two-photon annihilations of electron-positron pairs, theory [30] predicts polarization correlations, which can be described either as perpendicular correlations of linear polarizations of the two photons, or as both photons having the same helicity, if their circular polarization is analysed. The first type of correlation was observed in the azimuthal variation of the coincidence rate of Compton scattered quanta [31], evidence for the second type was claimed in an another experiment [32].

6.4 The determination of the helicity of the neutrino

In the beautiful experiment by M. Goldhaber et al. [33] on the helicity of the electron neutrino, its unknown helicity was transferred to the circular polarization of the subsequently emitted γ-ray. The first sentence of the paper gives all the important elements of physics involved. It reads 'A combined analysis of circular polarization and resonant scattering of γ-rays following orbital electron capture measures the helicity of the neutrino.' The sensitivity of Compton scattering to the state of circular polarization of the photons, when scattered against polarized electrons, means that the transmission rate of circularly polarized γ-radiation through a magnetized iron filter will depend on the direction of the magnetizing field relative to the line of flight of the photons. This was used in the neutrino helicity experiment to determine the photon polarization. The final result was that the neutrino has negative helicity and this in turn added strength to the V-A theory for weak interactions.

In this way the Klein-Nishina collaboration, and in particular the pioneer work by Nishina, came to play a role in the study of parity violations in weak processes and in determining the nature of the weak interactions.

6.5 Other applications

The results of the collaboration between Oskar Klein and Yoshio Nishina are of importance in many other cases within nuclear physics, particle physics cosmic rays physics and astrophysics. However, time and space does not allow an exhaustive review of all that.

Finally, let me mention that the Klein-Nishina formula still serves as a favoured example in university courses on quantum electrodynamics.

Acknowledgements

I am much indebted to professor Ryogo Kubo, the President of the Nishina Memorial Foundation for the invitation to give a paper on the collaboration between Klein and Nishina at the Yoshio Nishina Centennial Symposium, held in Tokyo on December 5-7, 1990. For useful and informative discussions I would like to express thanks to professor Chen Ning Yang, Stony Brook, USA, and to my colleagues in Stockholm, Prof. Bertel Laurent and Prof. Bengt Nagel. Discussions with docent F.D. Lars Bergström, Stockholm, have been of special importance, for which I would like to express my sincere thanks. I am most indebted to Dr Finn Aaserud, Copenhagen, for his kindness in searching the Niels Bohr archives for correspondence relating to the Klein-Nishina work, and to Dr. Tamaki and his assistants Dr Takeuchi and Dr Yazaki at the archive of the Nishina Memorial Foundation for help in finding relevant material in their files in Tokyo. Thanks go to Professor Michael Albrow for his reading of the manuscript.

References

1. Oskar Klein: *From a life of physics* (World Scientific, Singapore 1989) pp. 69-84 , and Curriculum vitæ by Klein.
2. O. Klein, Zeitschr. f. Physik **37**, 895 (1926) ; ibid. **46**, 188 (1927)
3. O. Klein: Nature **118**, 516 (1926)
4. Th. Kaluza: Sitzungsber. d. Berl. Akad., p 966 (1921)
5. O. Klein: Zeitschr. f. Phys. **41**, (1927), 407
6. Cf. e.g. A. Pais: *Subtle is the Lord*, (Oxford University Press 1982)
7. A. H. Compton: *X-Rays and Electrons*, 1st ed. (Van Nostrand Co. 1926) pp.296-314
8. P.A.M. Dirac: Proc. Roy. Soc. A **111**, 405 (1926)
9. W. Gordon: Zeitschr. f. Phys. **40**, 117 (1927)
10. O. Klein and Y. Nishina: Zeitschr. f. Phys. **52**, 853 (1929)
11. O. Klein and Y. Nishina: Nature **122**, 398 (1928)
12. I. Waller: Zeitschr. f. Phys. **61**, 837 (1930)
13. I. Tamm: Zeitschr. f. Phys. **62**, 545 (1930)
14. O. Klein: *To the Memory of Yoshio Nishina*, manuscript (1975), the archive of the Nishina Memorial Foundation, Tokyo, (Japanese transl.) Butsuri, vol.**45**, No 10, 720 (1990)
15. B. Laurent: *private communication*
16. O. Klein: letter to N. Bohr, July 6, 1928, The Niels Bohr Archive BSC (13,1)
17. Y. Nishina: letter to N. Bohr, August 8, 1928, The Niels Bohr Archive BSC (14,2)
18. Y. Nishina: Zeitschr. f. Phys. **52**, 869 (1930)
19. W. Franz: Ann. d. Phys. **33**, 47 (1938)

20. S.B. Gunst and L.A. Page: Phys. Rev. **92**, 970 (1953)
21. P.A.M. Dirac: letter to N.Bohr, November 26, 1929, quoted in ref. [22] p 350
22. A. Pais: *Inward Bound*, 1st ed. (Oxford University Press 1986)
23. Cf. e.g. J.D. Bjorken ans S.D. Drell: *Relativistic Quantum Mechanics* (McGraw-Hill 1964), pp 10-12.
24. P.A.M. Dirac: Proc. Cambr. Phil. Soc. **26**, 361 (1930)
25. G.T.P. Tarrant: Proc. Roy. Soc. A **139**, 345 (1930)
26. L. Meitner and H.H. Hupfeld: Naturwiss. **18**, 534 (1930) and Z. Phys. **67**, 147 (1931)
27. C. Y. Chao: Proc. Nat. Acad. of Sci. **16**, 431 (1930)
28. Cf. Bing An Li and C.N. Yang: Int. J. Mod. Phys. A **4**, 4325 (1989)
29. C. Y. Chao: Phys. Rev. **36**, 1519 (1930)
30. J.A. Wheeler: Ann. N.Y. Acad. Sci. **48**, 219 (1946)
    C.N. Yang: Phys. Rev. **77**, 242 (1950)
31. R.C. Hanna: Nature **162**, 332 (1948)
    E. Bleuler and H.L. Bradt: Phys. Rev. **73**, 1398 (1948)
    C.S. Wu and I Shaknov: Phys.Rev. **77**, 136 (1950)
32. F.P. Clay and F.L. Hereford: Phys. Rev. **85**, 675 (1952)
33. M. Goldhaber, L. Grodzins and A.W. Sunyar: Phys. Rev. **109**, 1015(1958)

Does Physics Ever Come to an End?

R. Peierls

Nuclear Physics Laboratory, Keble Road, Oxford OX1 3RH, UK

Abstract. The answer depends on the interpretation of "Physics". If the term is restricted to the search for the fundamental laws, the answer is "yes", there will be an end. If physics comprises the expansion of our understanding of the consequences of the laws of physics, there will be no end.

The subject of my talk seems to be the only one in this morning's session which does not explicitly relate to Nishina, but I do not want to miss the opportunity of expressing my great respect and admiration for him. I never had the privilege of meeting him, but I know enough about his contributions to physics and to the development of physics in Japan to be impressed and grateful.

The answer to the question I have been set, "Does physics ever come to an end?" depends, of course, on what we mean by physics, and on this the physics community, or even this audience, will not be unanimous. Some will see physics as the search for the basic laws of nature, and regard it as the end of physics if this search ceases. This is a view often held by specialists in particle physics, who sometimes show a rather arrogant attitude to other branches of our subject. Others feel that a much broader range of problems are good physics and worth pursuing.

So we have really two questions, and I shall start by discussing the one relating to the search for the fundamental laws. Talk of an impending end is not new. For example, in the early 1930s, when atomic theory was complete, there remained only some problems with relativistic field theory, and there were two dimensionless parameters: the fine structure constant, and the electron-proton mass ratio. These two numbers seemed to be related. There was also the nucleus, but so little was known about the nucleus that there did not seem many problems. Besides, the nucleus evidently contained electrons, and we could not understand that, but it was evidently related to the problems with relativistic electron dynamics. So all remaining problems seemed to be related, and some expected that they would be resolved in a single step. Of course we know today how wrong was this view.

For a more general analysis I cannot do better than quote the talk on the future of physics which Richard Feynman gave at the M.I.T. Centenary: He reminded us that the development of physics has proceeded in steps, each of which solved the problems at one level by opening up a new and deeper level, with new problems, below. The properties of chemical compounds were illuminated by introducing molecules and atoms; the problem of the properties of the atom were explained in terms of nuclei and electrons; the nucleus was recognized as consisting of nucleons; the forces between these were seen by Yukawa to be based on new fields, in particular that of pions, and that and the problems of other new particles led us to the picture of quarks and gluons. Trying to understand the weak interactions led us to another new level of problems. In the light of this, Feynman said, there are three alternatives for the future development:

Springer Proceedings in Physics, Volume 57
Evolutionary Trends in the Physical Sciences Eds.: M. Suzuki and R. Kubo

The first possibility is that we might one day find the ultimate basic laws. That, he said, would mean the end of physics. It would be sad, he added, because it would leave us defenceless against the philosophers, who always try to prove that things must be the way they are - so far we have always been able to disconcert them by finding new and unexpected things.

His second possibility was that, before we reach the basic laws, we might run out of facilities. There may be a limit to the energy that can be reached in practice by accelerators, and to the intensities of their beams. Within these limits we may not have enough clues to piece together the fundamental story. Then physics would also end.

Finally, he said, it was possible that such limitations could be overcome by new and ingenious methods of accelerating particles, or of making discoveries without raising the particle energy. Yet we might never reach the ultimate basis. In that case we would go on discovering layer after layer of fundamental physics, and then we would get bored. We could not sustain our excitement in such an endless succession of levels. He therefore concluded that, whatever happens, we live in the golden age of physics. I think it is hard to quarrel with this analysis. Feynman's second alternative, the limits on energy, has acquired more plausibility since his talk. Considering the cost of the next generation of accelerators, and the effort needed to get them funded, one feels there may not be many generations ahead.

As regards the search for the fundamental laws we must therefore answer the question in the affirmative. Physics will come to an end, unless you allow the rather gloomy kind of survival of Feynman's last alternative.

Feynman did not specify a time scale; in fact his talk was supposed to deal with physics for the next thousand years! There may well be some uncertainty at what stage we shall recognize which of Feynman's possibilities is realized. For example, the possibility of reaching the ultimate laws, so that fundamental physics would be complete, may well be controversial. Remember that, with the development of physics, also the range of questions which a physicist should ask, expanded. At the end of the last century, for example, Ohm's law was physics, but the value of the conductivity of copper was not. This was something you looked up in a table, or on which a chemist might have views.

Some will not be satisfied that physics is complete until many deep questions have been answered. Not only what are the ultimate constituents, what kind of particles or fields, or whatever new concepts we may meet, and their interactions, but also why just these particles and no others, why just those interactions, etc. Taking this questioning to its extreme would mean deriving all physics from a priori principles, and most of us would doubt that this is possible. So there may well be controversy whether physics is complete and whether the end has been reached.

Or consider the second possibility, that we run out of clues. Here of course there will always be theoreticians who refuse to give up, and who will try to construct theories from the available clues, reinforced by their ingenuity and imagination, much like the present fashions for strings and supersymmetry, though I leave you to decide whether this is physics or science fiction. Such approaches may succeed, so in time what appeared to be the second situation might become the first. This, of course, may also involve controversy. Since we assume there are no new experimental clues to be obtained, such theoretical laws may become a matter of faith. Dirac believed that a theory could become convincing because of its mathematical beauty, but estimates of beauty may not be unanimous.

In the third possibility, the infinite succession of levels, there would exist for a long time some physicists willing to plod on, and it is not clear when finally everybody would give up, or the community would discontinue the support of physics, but it would probably be within Feynman's thousand years. Eventually we are bound to find ourselves in one of the three situations outlined, and then the search for the basic laws of physics will come to an end.

Let us now turn to the other way of interpreting the question: including in physics all the beautiful work done in areas in which the laws of physics are well known. From a naive point of view, once the laws are known, all one has to do is to apply them to various situations, and activity that might be described as Applied Physics if it is not Engineering. But this is to misjudge the position completely.

There are indeed cases in which new and exciting devices are invented by making predictions from the laws of physics. Such an example is the laser. Its inventors saw the theoretical possibility of such a device, but this was no trivial matter; many people knew the laws of spectroscopy and had been applying them for many years. Yet nobody had seen the point. Of course seeing the possibility of the laser in principle was not the whole story. One had to know the spectra of various substances, to find combinations of levels and transitions suitable for the purpose, and then it needed experimental skill to implement it. Since then the development has continued, adding to the variety of lasers (including tunable ones) and also discovering many new ways in which lasers can be used. This is where the engineer comes in.

But this progress by theoretical prediction is by no means general. An almost complete opposite was the cause of superconductivity. It is not only that it was not predicted theoretically. After its surprise discovery we did not understand what was going on, and even after quantum mechanics had led to an understanding of most solid-state phenomena it remained a mystery, even though everybody was convinced it must be a consequence of the known laws. It was not until 1957 that the explanation was found by Bardeen, Cooper and Schrieffer. Their work built on many ideas that had been proposed to find an explanation, particularly by F. London and by Frohlich. The B.C.S. theory not only helped us to understand and therefore better to apply the phenomenon, but it also enlarged the scope of theory, since the techniques that were necessary for the explanation had important applications to other problems, including those of field theory.

Superconductivity has many important applications, and has in many instances been taken over by the engineers, but that is not always the case with exciting discoveries. The superfluiduity of liquid helium is another effect that came out of the blue, without being foreseen by theory, though Landau soon proposed a description which no doubt contains the essence of the phenomenon. The superfluidity of He3 also involved a surprise. Before its discovery many suggested that He3 would turn superfluid at low enough temperature; in fact many expected this would happen well before one reached millidegree temperatures. But the properties were unexpected. The analogy with superconductivity had suggested that the helium atoms are paired, and so they are, but not as expected in a singles-S state. Different pairing states make the phenomenology of superfluid He3 very rich.

Superfluidity of either He isotope has not so far found practical applications and is not likely to do so, but it is exciting physics. The lack of applications ensures that the physicist can keep the field to himself, without danger of being displaced by the engineer. This is not so with the current excitement over the high-temperature superconductivity, which also came as a complete surprise and has as yet no generally accepted explanation. Here the potential practical applications look so promising that there is a race between physicists and engineers. A colleague remarked that this is an ideal time to work on other problems in condensed-matter physics, because you have no competition - everybody is so busy working in high-temperature superconductivity. Perhaps this is a little exaggerated.

Another beautiful subject opened up in the post-war years is the study of phase transitions. For a long time it was taken for granted that in a second-order phase transition, such as the critical point of a gas-liquid system, or the Curie point of a ferromagnet, the specific heat was discontinuous but finite. In fact, Onsager had given a rigorous solution of the two-dimensional Ising model, in which the specific heat at the Curie point is singular, but this was not noticed. Only when precise measurements showed that the specific heat at the critical point gets bigger as the accuracy of the measurement increases, was it realized that the correct behaviour is singular, and from this the study of such transitions by means of the renormalisation group has developed.

As one of the latest surprises I might mention the quantum Hall effect. A two-dimensional electron system (e.g. confined to the surface of a semiconductor) shows qualitatively the same Hall effect as a metal or a semiconductor, but there are periodic variations, which follow simple numerical relationships with incredible accuracy. This, too, has called for novel and sophisticated theoretical approaches.

Astrophysics in a sense straddles the boundary between my two categories: much of it is covered by the well-established laws of physics, but some of its aspects, such as the solar-neutrino problem and that of the "dark matter" in the universe, involve the frontier of known laws.

I could not possibly list all the recent exciting developments. I talked about these few examples to illustrate, firstly, that there are non-trivial studies for which the skill, the experience and the imagination of the physicist are required. I would insist that they are certainly physics. Secondly, as the examples show, they arise unexpectedly, not from any systematic attempt to deduce consequences of the laws of physics. I see no reason why the flow of such discoveries should ever stop. I do not have the confidence of Feynman to make predictions for a thousand years, but if "physics" includes the type of phenomena I have described, there is certainly no end in sight.

So, depending on how you interpret the question, the answer is either "yes" or "no".

Part II

Scientific Session

Where Do We Go from Here?

L.M. Lederman

University of Chicago,
5640 Ellis Avenue, Chicago, IL 60637, USA and
Fermilab, P.O. Box 500, Batavia, IL 60510, USA

1. Where are We?

1.1 Introduction

Of course we are in Tokyo, celebrating the 100th anniversary of Yoshio Nishina, a scholar whose activities encompassed so many different fields.

Among the many accomplishments of Yoshio Nishina we must remember that accelerator science was one of his major interests and he directed one of the foremost accelerator labs in the world up until the war.

In this talk I will summarize where we are, emphasizing those aspects of both theory and experimental science which are likely, in my opinion, to be springboards into the future. Unlike Nishina, I will stick to high energy particle physics although the guidance and strong influence of cosmology must of course be included. If you notice that I spend more time on experimental facilities than on the prospects for superstring theory, it is only that I truly believe the road to the future as we now dimly see it, is more likely to require new machines and new detectors than improved mastery of Calabi-Yau manifolds. Of course, we are inherently guided by theory and where we are going will very likely have the same felicitous blend of theory and experiment as we enjoy now. One thing about the future compared to the present is that it is undoubtedly longer.

1.2 Theory

We begin our springboard survey with a reminder that we live in the shadow of an incomplete Standard Model. This teaches us that the matter in the world is made up of six quarks and six leptons. In each family there is a missing member. In both cases, the absent particle has a very special role and both particles, the as-yet- undiscovered top quark and the not-yet detected tau neutrino will in fact play prominent roles in our future. For now the puzzle has to do with why the top mass is so heavy, (it is at least 90 GeV according to Fermilab results) and whether the tau neutrino has any mass at all and if so, is it enough to make the expansion parameter of the universe W = 1?

Springer Proceedings in Physics, Volume 57
Evolutionary Trends in the Physical Sciences Eds.: M. Suzuki and R. Kubo

These mass puzzles extend over all the matter particles in the entire standard model and, are a 1990 version of Richard Feynman's 1950 question: "Why does the muon weigh?"

To complete my description of the Standard Model, the matter particles are beholden to the electroweak and the strong force. These are represented by 12 gauge bosons. Here too something is missing and again it is related to masses.

The unitarity crisis required the introduction of a new interaction carried by a neutral scalar particle, the Higgs. This field has the added feature of being capable of mass generation, giving the Z^0 a large mass and thereby breaking the symmetry in the electroweak interaction. It seems likely, if I understand what my theory colleagues are saying, that all fermion masses are generated as potential energy in the Higgs field. Well, if true, Higgs is crucial to any advance and we must try to find Higgs particles. The Higgs mass is an open parameter of the SM and here again we have an important (and very expensive!) springboard to where we are going. Fortunately there is a clue in that the theory becomes inconsistent (Higgs-Higgs scattering etc.) unless the mass of the Higgs is less than 1 TeV or so. This limit motivated the design of the SSC, the 40 TeV proton-proton collider now under construction in Texas.

I will select just one more of the questions left open by the SM and that has to do with CP violation, the ability of neutral K-mesons (the K-Long) to decay to 2 pions. This reaction has vast cosmological implications, nothing less than the "origin of matter." There has been a tremendous experimental effort to measure CP violating parameters and these will surely continue but the more recent possibility of studying CP violation in the B-meson ($b\bar{d}, b\bar{s}$) system has spurred proposals for the construction of machines specifically designed to do these things. These are usually called Beauty factories.

Finally, we must realize that the story of particle physics is a mixture of futures; futures motivated by theoretical crises and predictions and futures motivated by experimental and technological opportunities. For example, the $\pi, \bar{p}, \nu_\mu, K^\circ_L, W, Z,$ predictions lead to searches, new accelerators and new techniques, the $\mu, K^\pm, \Lambda^\circ, CP, J/\psi, {}^\circ, \tau$ were surprises, gifts of new techniques and of machines. If history is a guide, we will use our increasing powers of observation and measurement to test today's theory but also to search for new phenomena.

1.3 Facilities

The inventory of front-line machines is a decreasing function of time. In 1990, the Fermilab Tevatron provided nearly 2 TeV for $p\bar{p}$ collisions at a luminosity which permits observation of 10^5 collisions per second. It is likely that this will be increased by a factor of 50 or so by the mid-1990's. Such a luminosity ($10^{32}cm^{-2}sec^{-1}$) could permit the observation of processes that have cross-sections as small as $10^{-37}cm^2$. The Fermilab collider is the only existing machine that can produce the top quark. The Fermilab fixed target program at 800-900 GeV also provides the highest energy collisions of a wide variety of primary, secondary and tertiary particles. If we are to see the tau neutrino, it will almost certainly be in the fixed target program at Fermilab

At CERN there is a $p\bar{p}$ collider which pioneered the technique of creating intense antiproton sources and then head-on collisions of protons and antiprotons. This machine produced spectacular data: discovery of W and Z as well as "jets." However with an energy of 630 GeV, it is scheduled to close in 1991.

The CERN LEP machine, currently running at 50 GeV e^+ colliding with 50 GeV e^-, is a Z^o factory.

The scheme of this 27 km circumference machine with its four large and sophisticated detectors is to study the decay modes of the Z as a probe of new physics and as a means of establishing important SM parameters with great precision. They are approaching 10^6 Z^o's and in a few years, perhaps as many as 10^7 Z^o's so that very rare decay processes can be seen. Also in the next few years this machine will go to a total energy of about 180 GeV in order to study WW, ZZ, WZ, and γW pair production processes.

The SLC machine at SLAC manages the same collisions as LEP but in an accelerator of innovative design using the SLAC linac to accelerate e^+ and e^- and bring them together in two semicircular tracks. The machine is an approach to a linear collider, a much studied configuration for producing much higher energy e^+e^- head-on collisions. Unfortunately, its luminosity is only a few percent of LEP.

The smaller e^+e^- machines at Cornell (CESR) KEK, DESY, SLAC are providing detailed data on SM properties with the CESR and DESY machines until now providing the bulk of the data on B^o mesons. Lower energy fixed target machines at BNL and CERN have very selective programs e.g. Brookhaven's study of very rare K-decays and CERN's precision measurements of CP violation and of neutrino scattering. The Beijing charm factory has recently entered the field and will continue the work carried out at SLAC's SPEAR. HERA, a unique e p collider (30 GeV e's x 800 GeV p's) is scheduled to turn on in 1991 and will provide both search and measurement data.

Finally, I would tell you a bit about the apparatus. We are today in a situation where groups of 200-500 physicists can, in 6-8 years, assemble collider detectors of impressive complexity, making use of data acquisition systems and computational power that rival the accelerators in cost and technical sophistication. Consider the CDF dectector at Fermilab. It looks at 10^5 (soon to be over 10^6) events per second, each with up to 100 tracks and about 10^4 bytes per track. This is 10^{11} bytes per second. An on-line system examines these events and by a process of sequential filtering, finally writes about 5 events per second to tape. This is the springboard to the supercollider or CERN's version, the LHC, where the problem grows to 10^{15} bytes per second!

We have no time to describe the quality of the data, the trajectory measurements, the precision track-origin locators (to $\pm 10\mu$), calorimetric energy measurements, etc.

2. Where Are We Going, (Part A)?

Let's review the selected SM weaknesses in order to trace these threads into the future. We discuss these in the context of presently available accelerators.

2.1 Top Quark

We already know that $M_t > 90$ GeV. My own puzzlement is illustrated by a new table of the Standard Model which I call the Lego SM plot (see Fig. 1). The diagram is designed to emphasize the puzzle of the massiveness of the top quark. The sensitivity of the search for the top quark depends on the energy of the colliding quarks (partons) and on the integrated number of collisions. The above limit was based upon about 10^{11} collisions or an integrated luminosity of 4.2 pb^{-1}. In the 1991 run of the TEVATRON collider, the CDF detectors will be joined by a new detector, DZERO. It is expected that each detector in the 1991 run will have an integrated luminosity of 20-30 pb^{-1} which enables the mass range of up to about 130 GeV to be searched. By 1996, given the upgrades Fermilab has proposed, the top will be found if its mass is < 250 GeV.

Theoretical consistency of data on B mesons, on the W mass (within the SM) leads to the conclusion that $M_t < 250$ GeV. This is because the top quark enters in radiative corrections to SM parameters. If the TEVATRON does not find the top quark, the SM is incorrect. (The Higgs thing also enters into this argument). The issue in the quest for the top quark is then to know the mass and to determine whether the huge mass is merely an accident or is it some signal (see Fig. 1) that top is special and its properties will tell us about the very nature of mass.

2.2 Beauty Meson Factories

We mentioned that all the data on B's comes from the e^+e^- machines. Although the hadronic production of b-quarks has a much greater cross-section, until very recently backgrounds have prevented competition. However, excellent mass resolution has enabled CDF to reconstruct B^o events and study the specific mode:

$B \rightarrow J/\Psi + K$

It is expected that the next CDF run which will have a silicon vertex detector should collect about 100 times the number of B events. However, there is now a world-wide effort to design an e^+e^- beauty factory with work going on at KEK, SLAC, CERN, SIN, and NOVOSIBIRSK. The motivation is CP violation which promises to be very informative if seen in the $B^o\ \bar{B}^o$ pairs. B-factories are designed so that they can measure CP violation in a year's run.

Hadron machines hope to get in the game. The ratio of B production to total cross-section is only 10^{-6} (fixed target) or 10^{-3} (collider). The CDF B signal now has as many reconstructed B's as do the e^+e^- colliders. The evolving technology and ingenuity may well make this an interesting race, i.e. between existing hadron machines i.e. FNAL's collider and fixed target vs the e^+e^- machines, existing and proposed.

2.3 Neutrinos

Since Pauli's inspired speculation, neutrinos have continued to puzzle and lead physics to new ideas. Try to explain to a science writer that there is a particle that has no charge no radius and no mass but that it enables the sun to shine, to cool stars, and to distribute the heavy elements cooked in dying stars, throughout the universe! No mass?

The limit on m_e is about 10 ev, on ν_μ it is 200 KeV, the tau neutrino can be as heavy as 35 MeV.

The neutrino structure and especially the possibility of finite mass is one of the outstanding problems today and clearly a springboard to major research over the next decade. The current research had three motivations: (i) The famous solar neutrino problem (what depletes the flux of ν_e's?) (2) The dark matter problem, i.e. we need weakly interacting neutral particles with some mass (not too much!) and neutrinos are good candidates because they do exist; and (3) the width of the Z insists that a fourth generation neutrino, if it exists, must have a mass greater than 40 GeV.

A vigorous use of neutrinos as tools for studies of quark structures and weak interactions led to detectors of 1000 tons. The proposals now emerging involve higher intensity neutrino beams e.g. the Fermilab Main Injector machine which would increase the collider luminosity would also yield 10^{13} protons per second at 120 GeV and a superintense neutrino beam. They also involve more sensitive searches for neutrino oscillations and for the detection of the tau neutrino. Since we know least about ν_τ, it has been considered the most likely candidate for astrophysical dark matter. The question of whether the τ-neutrino has mass is crucial here. If it does have a mass, the mechanism that generates it is "....a window on the world beyond the SM." Some proposers insist that neutrino beams be aimed at detectors hundreds of kilometers away (long baseline oscillations).

Finally we should mention neutrino astronomy and solar neutrinos. We know there is an ambient flux of neutrinos from outside our solar system and even outside our galaxy. The detection of some 11 events from SN1987A in Toyama and Cleveland marked the first time non-electromagnetic signals have been received from outside the galaxy. Since γ -rays of PeV (10^{15}eV) have been detected, since these are generated by hadrons, these must almost certainly also generate neutrinos via hadronic weak interactions. Detecting TeV neutrinos would be a cosmological bonanza. The subtleties of solar neutrinos may indicate oscillations generated by fractions of an eV mass difficiencies between neutrino species. These in turn could have a vast influence on the large scale structure of the universe.

In summary, we touch the problem of mass again with neutrinos since it is not easy or natural in the SM to generate mass for neutrinos.

Thus oscillations or any direct way of observing ν–mass must require theoretical extensions beyond the Standard Model.

2.4 Higgs

We noted that the Higgs particle mass is an open parameter which can be as high as ~ 1 TeV. There are some theoretical estimates based upon an idea of Nambu by several authors [1] which is inspired by the massiveness of the top. These theorists attempt to make the "Higgs" a bound state of top and antitop. These models give specific predictions for the masses of the top and the Higgs, in the domain of 100-200 GeV.

Whether or not the $p\bar{p}$ machine can find a 200 GeV Higgs is an open question and depends critically on the luminosity of the improved TEVATRON.

3. Where Are We Going; (Part B)

This history has gotten off to a lively start. SSC was "conceived" in the late 1970 ICFA studies but it was brought to a sharp focus as a national plan in 1982. By July, 1983, it was embraced by the DOE and there began a serious design study under M. Tigner at the LBL headquarters of the SSC Design Group. The energy is 20 TeV in each beam yielding a splendidly violent 40 TeV in the CM with a collision rate of 10^8/sec.

Magnet R&D aimed at SSC was diversified to three laboratories (LBL, FNAL, BNL) and did not break speed records.

In 1987, SSC became U.S. policy, the site was selected and the SSC Laboratory founded in Texas under Roy Schwitters. As of current writing, the cost estimate for the SSC "hovers~ between $7.8 billion and $8.3 billion."

So what is the scientific drive for SSC?

We start with the list that any Congressman is completely familiar with:

1. Higgs! Electroweak symmetry breaking and SM predict that the reaction: $H^o \to Z^oZ^o \to$ 4 leptons will be seen at SSC if the mass is less than 800 GeV.
2. Z's, W's, copious production in pairs
3. Top physics
4. SUSY searches
5. Compositeness, is the quark (electron) a point?
6. Strong W_LW_L scattering
7. B physics
8. New physics which "explains" CP, 3 generations, quark lepton symmetry etc.

There is an incredible literature on the physics at SSC, and/or its European version, LHC. At this time, "expressions of interest" running to hundreds of pages have been received. This confirms the notion that interest is worldwide. About 5 or 6 propose to build generic detectors which are modelled on the DO and CDF or UA 1, 2 style of "4π" do-everything detectors. One detector proposed by 837.5 authors is a 6π detector. Other expressions of interest vary from the ubiquitous logs physics, to fixed-target beauty research. So far, only one, perhaps two, seem to be based upon totally new technologies. The next few years will see a refinement of these expressions-of-interest.

One of the more challenging aspects of SSC experimentation has to do with the collision rate. The design luminosity would yield 10^8 interactions/sec, each interaction generates ~100 particles requiring ~10^7 bytes to describe. This data rate requires all kinds of new techniques, radiation-hard detectors and up-close electronics, a refined mechanism for selecting the interesting events, etc. Whereas very few experts would claim that this problem is now completely solved, there is nevertheless considerable pressure to go to 10 times this rate or even more! From the theoretical physics point of view it is clear that this would help the Higgs problem, but from the experimental point of view, it is not at all clear that 1990-1992 technology can deal with these kinds of data rates.

4. Where Are We Going; (Part C: Beyond SSC)

It is the 130th anniversary of Yoshio Nishina. The year is 2020.
So by now we can also invent SSC results, e.g.

H_1^0 = 422 GeV found at SSC in 2004

H_2^0 ≅ 699 GeV but only 3s

Indications exist that there is a Higgs sector with a rich Higgsian spectroscopy.
To study this, we obviously need higher energy.

SSC may instead discover a new class of strong interactions which may, in the words of Steven Weinberg, revive the physics of our youth; dispersion relations, Regge poles, sum rules, all at a much higher energy. Again we'll need a machine appropriate to the energy. To decide the state of hadron colliders, we are fortunate to have the well-tested Livingston Chart (Fig. 2). This predicts that by 2030, we will have 1000 TeV in the CM. In order not to violate this schedule, we must start in 2020. The dilemma facing us in 1990 is that we can't know now what kind of facility will be appropriate. Of course by 2020, we'll know!

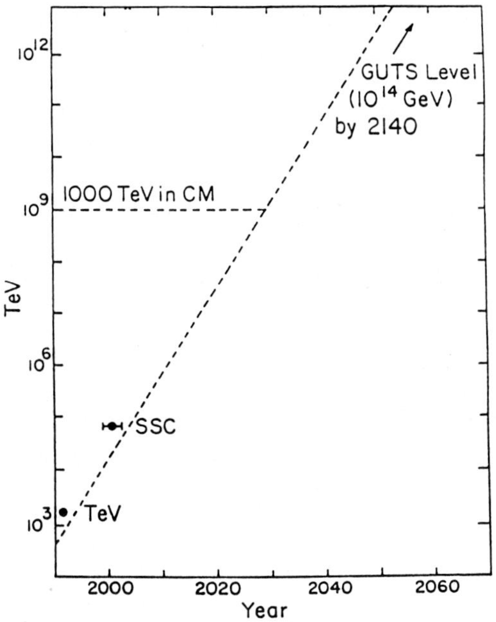

Figure 2. Upper reaches of the Livingston Plot whose absolute validity is established in the off-side years 1930-1990.

4.1 Electrons vs Hadrons?

There is a segment of devotees of e^+e^- collisions that seem to hold to a belief that the next machine after SSC "belongs" to electrons and this is as sensible as if the experts on Geiger counters would insist that they be employed on the next detector. The point is that we are all driven by physics. Electron machines were powerful in the 1970's and LEP's contribution to Z° physics, especially the width, is clear. The virtue of electrons, their clean initial state, may however count for less and less as the violence increases. Very narrow resonances like the Z^0, strongly coupled to electromagnetism, is one of the few states that strongly favor e^+e^- machines and these may be a vanishing breed at post-SSC energies. If hadron colliders can solve the rate problems and the messiness of the spectator partons, its relative economy in dollars per GeV and its large variety of initial states may win over e^+e^- colliders in the next round. A strong indication of this does not have to wait for SSC results in the 2000's but will be guided by 10^{32} luminosity in the upgraded Tevatron in the mid-1990's. If constituent collisions continue to be as clearly discernible at these rates, it will be a strong indicator that a 500 TeV x 500 TeV pp machine can be the 2020 machine, rather than the equivalent 50 TeV x 50 TeV e^+e^-. We must keep our minds open and weigh the physics potential of these two approaches. Both have formidable challenges, the former is largely in cost reduction. In the e^+e^- case, the technical challenges are so daunting that it is likely that the only sensible approach is an iterative, learning process, through a, e.g., 200 GeV x 200 GeV collider, then a 1 TeV x 1 TeV, etc. Each process is in the billion dollar category and probably requires of the order of ten or more years. Thus some imaginative efforts at magnet R&D to reduce costs of the post-SSC accelerator should start in the period of 1995-2005. Progress in high temperature superconductors is clearly relevant.

A design of a 500 TeV x 500 TeV machine was carried out in 1985 by J.D. Bjorken. The only daunting problem was the cost.

Some speculative theoretical ideas [2] in fact would strongly favor hadron accelerators in the hundreds of TeV range. These ideas are related to the notion that electroweak interactions become strong (non-perturbative) at high energies. Violations of B (baryon number) and L (lepton number) could be induced by new gauge fields (instantons). Observations of large probabilities of huge multiplicities in quark-quark collisions are possible outcomes of these ideas. What is involved is nothing less than the topological structure of the electroweak vacuum. So there! Both theoretical and experimental progress is needed before using these ideas as a decisive issue in this mythological next accelerator. However it does support the thesis that it is not at all certain that this will be an electron linear collider. It should be noted that in Europe, the "Eloisatron" concept of a multi hundred TeV hadron collider has been discussed by some of the more imaginative physicists for some years.

5. References

1. Y. Nambu, EFI 88-39, 1988
   Bardeen, Hill and Lindner, Phys Rev. D41, 1647 (1990
2. G. t'Hooft, Phys Rev. D14 (1976) 3432
   A. Ringwald, Nucl Phys B 330 (1990) 1
   S. S. Gerstein, Yu. F. Pirogov "Workshop on Physics at Future Accelerators" CERN 87-07 (1987)

Dynamical Symmetry Breaking*

Y. Nambu

Enrico Fermi Institute and Department of Physics,
University of Chicago, Chicago, IL 60637, USA

ABSTRACT

An overview is given concerning the concept of dynamical symmetry breaking and its examples in condensed matter, nuclear, and particle physics, including some speculations about the nature of the Higgs field in the Standard Model of electroweak unification.

1. Introduction

In my student days, Yoshio Nishina was one of those exalted names we talked about in awe. He was at that time heading cosmic ray and nuclear physics groups at Riken, the famous Institute for Physical and Chemical Research, which had played a unique role in the development of science and industry in Japan during the period between the two world wars. Having few, if any, professors in my university of Tokyo to teach us particle physics, we students used to frequent the weekly Riken seminars run by Nishina and his theoretical colleague S. Tomonaga. It was in this way that I was initiated into cosmic ray physics; I learned, for example, how Nishina's group was engaged in measuring cosmic ray intensities underground and over the Pacific Ocean. I also learned at first hand the mode of operation of the great school of theorists of the time, represented by people like Tomonaga, Yukawa, and Sakata, as they were developing their ideas about the cosmic ray mesons. I remember Tomonaga, at one of those seminars, reading Sakata's communication to him, in which Sakata was proposing that the cosmic ray "meson" (now called muon) and Yukawa's meson (now called pion) were different particles. However, this paper will concern certain theoretical ideas which have little direct connection with Nishina.

*Work supported in part by the NSF; PHY 90 0086

Springer Proceedings in Physics, Volume 57
Evolutionary Trends in the Physical Sciences Eds.: M. Suzuki and R. Kubo

2. Symmetry and Symmetry Breaking

The symmetry principle occupies an important place in our pursuit of physical laws, but it is not my intention to give an exhaustive discussion of the symmetry principle in general. Besides, my talk will inevitably reflect the fact that I am a particle theorist. When Yukawa created a theoretical paradigm with his meson theory in his search for the origins of nuclear forces, the concept of symmetry did not play any role. His paradigm, which I would like to call the Yukawa mode [1], was to hypothesize that behind new phenomena there are new particles in terms of which one can explain the former; the pursuit of particle physics is the pursuit of new particles. When a subfield of physics like particle physics was in its exploratory stage, this turned out to be a highly effective methodology. In fact it has remained so up to the present. But in the meantime the symmetry principle has also proven its power and importance as the field matured. In recent years we have seen a gradual emergence and even dominance of what I call the Einstein mode, in which theoretical principles drive the direction of particle physics. A key elemnt of this mode is the symmetry principle.

The purpose of this paper is to address one particular aspect of the symmetry principle, namely the dynamical, or spontaneous, breaking of symmetries. But the symmetry principle as it appears in modern physics has many facets, some of which had not been recognized before the recent developments in quantum field theory. So it seems appropriate for me to first give a brief summary of these various facets.

a) Symmetry gives a sense of esthetic beauty to physics and the natural world it describes. In mathematical terms, a symmetry essentially means a group of congruent operations under which the laws of physics are unchanged. The group may be continuous or discrete, and implies an associated conservation law which is respectively additive or multiplicative. Finding patterns of symmetry is highly useful in discovering regularities and conservation laws; conversely symmetry serves as a guiding principle in our search for a unified description of physical laws.

b) There are global symmetries as well as local, or gauged, symmetries. Wigner once remarked, according to my recollection, that there are two ways of establishing conservation laws: one by finding selection rules that apply between the initial and final states of a process, and the other by directly measuring the conserved charges by the fields they produce. Certainly this is a good characterization of the distinction between global and gauged symmetries. A gauged symmetry is richer and more restrictive than a global one in the sense that the former is in fact an infinite product of symmetries referring to each point of space-time. The Einstein gravity also belongs to this category. (Among the global symmetries one may include the so-called dynamical symmetries, like those found in the Keplerian and the harmonic motion, which are symmetries in the phase space, and are outside of the Noether theorem.)

c) It is often emphasized that physics consists of physical laws in local and differential form, plus the boundary and initial conditions which are subject to independent physical considerations. Symmetries usually refer to the former, but not necessarily to the latter. However, there are cases in which the topology of the physical space is coupled to that of the group manifold in question, so the boundary and initial conditions become an integral part of the symmetry. Topological considerations have led to concepts like solitons, monopoles, strings and instantons.

d) A symmetry implies degeneracy. In general there are multiplets of equivalent states related to each other by congruence operations. They can be distinguished only relative to a weakly coupled external environment which breaks the symmetry. Local gauged symmetries, however, cannot be broken this way because such an external environment is not allowed (a superselection rule), so all states are singlets, i. e., the multiplicities are not observable except possibly for their global part.

e) In reality global symmetries may be perfect or only approximate, leading to strict or approximate conservation laws. There may be a hierarchy of approximate symmetries, and often the patterns with which symmetries are broken are as meaningful and pleasing as the symmetries themselves. A symmetry may be so blatantly violated that it is the asymmetry rather than the symmetry that is interesting and significant. Parity violation in weak processes is an example.

f) There can be clashes of symmetries: different interactions may have different symmetries which are in conflict, and this conflict becomes the prominent feature of certain phenomena, again as in the case of the weak vs. the strong interactions, where their symmetry axes are tilted with respect to each other, so to speak.

g) A symmetry and the associated conservation law that are strict in classical theory may be violated by a quantum anomaly, i. e., the symmetry in question may be valid only "on shell", but not in the entire function space of fields over which the quantum action is defined. The chiral anomaly is a prominent example of it. Absence of anomalies is thought to be a necessary condition for a renormalizable theory based on gauged symmetries.

h) Symmetries inherent in the physical laws may be dynamically and spontaneously broken, i.e., they may not manifest themselves in the actual phenomena. The rest of the paper will address this topic in more detail.

3. Dynamical (Spontaneous) Symmetry Breaking

The fact that crystals, molecules and atoms exhibit symmetries as well as asymmetries seems to have caught the attention of physicists already when the group theory was being developed by mathematicians in the last century. According to Radicati [2], Pierre Curie [3] was one of the first physicists to discuss the aspects of symmetries and asymmetries in a modern language, mostly in the properties of crystals and of their responses to external forces. Be that as it may, a comprehensive historical review is not intended here.

The spontaneous breakdown of symmetries as a general concept is of more recent origin, although it predates the term coined by Baker and Glashow [4] in the 60's. The name is too long, and does not represent its content very adequately, but it has stuck for lack of a better one. It also appears that there exist subtle nuances in the way different people understand its meaning. Sometimes the term dynamical symmetry breaking is used as opposed to spontaneous symmetry breaking to denote dynamical mechanisms which are not immediately apparent. But in my opinion such a distinction is irrelevant. It is always a dynamical question whether a symmetry breaks or not. The two terms may be used interchangeably. Each term has its merits, but I will mainly use the word dynamical, and furthermore give it a rather narrow meaning. It does not include asymmetries of small finite systems like molecules (the Jahn-Teller effect). I will start with my definition of dynamical symmetry breaking, relying on concepts taken from group theory, statistical mechanics and quantum field theory.

As already mentioned, a symmetry implies degeneracy of energy eigenstates. Each multiplet of states forms a representation of a symmetry group G. Each member of a multiple is labeled by a set of quantum numbers for which one may use the generators and Casimir invariants of the chain of subgroups, or else some observables which form a representation of G. It is a dynamical question whether or not the ground state, or the most stable state, is a singlet, the most symmetrical one.

Consider now a system with a large number of degrees of freedom N, and the ground state is either degenerate or asymptotically degenerate so that its multiplicity grows and the energy splittings go to zero with increasing N. Usually one has in mind a uniform medium, where N is proportional to the number of constituents, and the spatial extent of the medium also grows with N. In the limit $N \to \infty$ (the thermodynamic limit) one may choose any particular state belonging to the degenerate multiplet, and call it the ground state of the medium. The quantum numbers of the state are infinite, but one may define their densities per unit volume, and call them order parameters.

Physical phenomena that can happen in this medium span a Hilbert space of states including the ground state under consideration. This space, however, is only a subspace of the Hilbert space of the system one had when N was finite. This is because the other ground states cannot be reached from the present one by means of local perturbations that operate only on a subset of its constituents. The two ground states are infinitely orthogonal, so to speak. The effective Hilbert space is one built on the present ground state by exciting it by local perturbations only. The system behaves as if the ground state was nondegenerate, but had reduced symmetry. Its symmetry (if one remains) is that of the subgroup H of G that leaves the order parameters invariant. The order parameters as a representation of G then belong to the coset space G/H.

According to the above characterization, the emergence of a superselection rule that reduces the Hilbert space is the essence of dynamical breaking of a symmetry as I would like to define it. It is crucial that N goes to infinity, but the symmetry may only be asymptotic, and the degeneracy need not be infinite. The familiar example of a double-well potential density [5] [7]

$$V(\phi) = G^2(\phi^2 - v^2)^2 \qquad (1)$$

for a real scalar field $\phi(x)$ has two minima $\phi = \pm v$. If the number N of points in space is considered to be finite, the twofold degeneracy will be lifted by tunneling. For an asymmetric state centered around one of the two minima, the kinetic energy that causes tunneling is a symmetry-restoring agent, but it becomes ineffective as $N \to \infty$, so no mixing will take place between the two degenerate states.

In the case of a continuous symmetry, the large-N limiting behavior becomes more subtle. If the real field ϕ in the above example is replaced by a complex one, one has a $U(1)$ symmetry. The order parameter $< \phi >$ then is determined only up to an arbitrary phase angle $\theta(mod\ 2\pi)$ that labels the degenerate vacua. Two vacua corresponding to two distinct θ's are orthogonal, but the phase θ may be regarded a field, and local variations of θ from the given constant value will generate excited states. If the region of variation becomes large and the wave length of its Fourier transform also becomes proportionately large, one approaches a constant (and nonlocalized) variation, which amounts to a transition to a different vacuum, hence no change in energy. From this argument one infers that there will be a normal mode, the Goldstone mode, of excited states which have no energy gap in the long wavelength limit. In relativistic theories, the Goldstone mode behaves as a relativistic massless particle.

The above statements about the existence of symmetry breaking and associated gapless modes have some exceptions. Basically it has to do with the effectiveness of symmetry-restoring forces, i. e., how big the barrier is between broken symmetry configurations. In the case of continuous symmetry, there is no potential barrier, only a kinetic barrier. As a result, the Goldstone mode can exhibit an infrared instability, i. e., its large wavelength zero-point fluctuations wash out the order parameters and restore a single symmetric ground state. This can happen in low-dimensional media.

Another notable exception is when the symmetry is a gauged one. If the complex field ϕ in the above example is coupled to a $U(1)$ gauge field, the phase θ is a gauge parameter. Fixing it to a constant breaks gauge invariance. As θ turns into the dynamical Goldstone field, it couples to the gauge field which is also gapless. The mixing of two gapless modes then lifts their degeneracy, and gives rise to a massive mode with three polarizations, the plasmon mode. One may also say that the gauge field causes long range correlations between constituents so one cannot gently modulate the order parameter; such a modulation gets shielded.

Often quoted examples of dynamical symmetry breaking are ferromagnetism, crystal formation, and superconductivity. In an isotropic Heisenberg ferromagnet, the total spin is conserved. Dynamics favors neighboring spins to be parallel, so the ground state of the system has maximum spin pointed in some direction. The symmetry breaks from $SU(2)$ (or $O(3)$) to $U(1)$ (or $SO(2)$), the latter being the rotation group around the chosen axis. Which axis the system chooses depends on the initial and boundary conditions or on the environment. A typical procedure is to impose a weak magnetic field which then is gradually switched off. The Goldstone mode is the spin wave (polarization perpendicular to the magnetization axis) belonging to the coset Lie algebra $o(3)/o(2)$.

A crystal is said to violate the Euclidean incariance $O(3) \times T(3)$ down to a discrete subgroup, i. e., the space group of the crystal, because one imagines it to be fixed in space. The kinetic energy of the center of mass motion in the Hamiltonian of the crystal is a symmetry recovering agent, but it vanishes in the infinite mass limit, so one can localize the system by an infinitesimam force, breaking momentum and angular momentum conservation. Because the Euclidean group is a semidirect product, the Goldstone modes corresponding to o(3) and t(3) are coupled, and one ends up having only three modes, the isotropic longitudinal and transverse sound waves in the long wavelength limit.

Superconductivity, as described by the Bardeen, Cooper and Schrieffer (BCS) theory [6] as well as by its predecessor, the Ginsburg-Landau (GL) theory [7], is a nontrivial example of spontaneous symmetry breaking. The BCS theory is a microscopic, and hence more fundamental, description than the GL theory which is a phenomenological representation of the former. But both have served as the prototype of theories for various phenomena in condensed matter, nuclear and particle physics. The essence of the BCS theory is the Cooper pair formation: the pairing of an indefinite number of electrons of opposite spin near the Fermi surface due to a phonon-induced attraction. It leads to a nonzero pair correlation function $< \psi_{up}\psi_{dn} >$, and analog of the magnetization in ferromagnets. This complex pair field carries electric charge, and its phase is the gauge parameter.

The notion of the dynamical symmetry breaking with its characteristic properties as defined above first emerged in an attempt to resolve the question of gauge invariance in the BCS theory [8]. The concept of degenerate vacuua and the analogy to ferromagnetism had also been invoked in an earlier work of Heisenberg [9]. In his nonlinear theory of elementary particles, it was assumed that some internal quantum numbers like isospin and strangeness were not the intrinsic attributes of the elementary fermion field, but were spurions (a sort of nondynamical Goldstone mode with zero momentum and energy) picked up by particles from degenerate state vectors of the world acting as a reservoir.

4. The BCS Mechanism

By BCS mechanism I mean here the formation of a Cooper pair condensate as an order parameter, due to an attractive interaction between fermions, typically a short range one. Some salient features of the mechanism are [10, 11]:

a) There are fermionic and bosonic excitation modes. The order parameter causes mixing of fermions of opposite charges (particle and hole) leading to an energy gap in the dispersion relation. The bosons are collective states of fermion pairs, and come in two kinds, the "π" or "Goldstone" mode, and the "σ" or "Higgs" mode, corresponding respectively to modulations of the phase and the modulus of the pair field.

b) There are two energy scales: that of the dynamics of the constituents and that of the energy gap which is usually lower than the first. The latter is dynamically determined by a gap equation as a nonperturbative solution. In the short range and weak interaction limit, the fermion and boson modes satisfy simple mass relations:

$$m_\sigma : m_f : m_\pi = 2 : 1 : 0, \tag{2a}$$

$$m_1^2 + m_2^2 = 4m_f^2. \tag{2b}$$

The second relation, of which the first is a special case, applies to a pair of extra bosonic modes (m_1 and m_2) that exist in p-wave pairs like in ^{3}He.

c) There are induced interactions among those modes. They are controlled by a single coupling parameter which represents the ratio of the gap and the constituent energy scale. One can therefore translate the low energy contents of the BCS mechanism into an equivalent and restricted Ginzburg-Landau-Gell-Mann-Lévy system [12] which contain phenomenological fermion and complex boson ("Higgs") fields, one dimensionless coupling, and one mass scale (the vacuum expectation value $v = < \sigma >$ of the "Higgs" field σ).

d) There exists what I call quasi-supersymmetry [13], of which the mass formula (2a) is a consequence. It means that the static part of the GL Hamiltonian can be factorized in terms of fermionic composite operators as in supersymmetric quantum mechanics and quantum field theory. More specifically, these operators are spatial integrals of the densities

$$Q = \Pi\psi + W(\phi)\psi^\dagger, \text{ and its Hermitian conjugate } Q^\dagger . \tag{3}$$

Here Π and ϕ are canonical conjugates expressed as $n \times n$ matrix fields multiplying an n-component fermion field ψ; W is the square root of the Higgs potential: $V = tr\ W^2$. These fermionic operators give rise to a spectrum generating superalgebra. The physical origin of the quasi-supersymmetry underlying the BCS mechanism is not clear, but it is possible to generalize quasi-supersymmetry to a relativistic quantum field theory in which the Poincaré part of the superalgebra can be realized among a set of fermion, Higgs, and gauge fields.

Among examples of the BCS mechanism are superconductivity, superfluidity in ^{3}He, and the nucleon pairing effects in nuclei. Bosonic modes satisfying the mass relations of Eq. (2) have been found in superconductors and ^{3}He [14].

As for the nucleon pairing, I have recently claimed [15, 11] that the Interacting Boson Model [16] of nuclear excitations may be interpreted basically as a GL description of the BCS mechanism at work in nuclei. There is a caveat to be made here, however. Nuclei are finite systems so the concept of spontaneous symmetry breaking does not apply in the literal sense, but the latter may nevertheless offer a reasonably good picture of the dynamics involved. The near degeneracy of a multiplet of nucleon valence shell states dictates the corresponding degeneracy of bosonic pair states of spin 0 and 2, and forms the basis of a GL Hamiltonian in terms of nucleon and boson fields. In a typical example, a $U(1) \times SU(4)(\sim O(6))$ symmetry of the six complex bosons is broken spontaneously to $Sp(2)(\sim O(5))$ after forming a condensate in one of the pairs states, and thereby breaks the baryon number $U(1)$. To take care of the fact that finite nuclei do not actually break $U(1)$, one projects the broken symmetry states onto unbroken ones, thus eliminating spurious modes. The boson-boson interaction obtained in this way reproduces the corresponding part of the phenomenological formula of the IBM fairly well in terms of the density and volume of the nucleus only.

In particle physics, the chiral dynamics of hadrons consisting of massive quarks, pion, sigma meson and others, is generally interpreted as a realization of the BCS mechanism in QCD, although not of the short range and weak coupling variety of the previous examples. Each massless quark field has a chiral (γ_5) invariance. The masses of quarks generated spontaneously by gluonic interactions are usually referred to as "constituent masses". The pion, which is pseudoscalar, is essentially the accompanying Goldstone boson, but it is not strictly massless because the chiral invariance is broken by small "current masses" already present due to electroweak interactions. Furthermore, chiral symmetry is in general anomalous, so one does not expect massless or nearly massless bosons to exist except for a particular linear combination of chiral transformations for which the anomalies cancel.

In a similar fashion, it is often thought that the Standard Model of electroweak unification may in fact be the low energy effective form of a more fundamental dynamical theory in which the Higgs field is a composite object. Recently I have suggested that the Higgs field is not formed out of new heavy fermions as in the technicolor theory, but rather of the top and antitop quarks. This will be discussed below.

5. Tumbling and Bootstrap

I now come to bring up some new theoretical possibilities related to spontaneous symmetry breaking. One is known by the name tumbling [17]; I will call the other bootstrap.

In the BCS mechanism a massless fermion field acquires a mass, and composite bosons are created at the same time. Consider a set of fundamental fields having chiral invariance, for example in a grand unified theory, which is valid at a large energy scale E_1. Suppose the chiral symmetry is broken, and a mass scale $E_2 < E_1$ is created. The various composite bosons can be exchanged (in the "t-channel") between the fermions. If the induced interaction is attractive, it may trigger a second round of Cooper pair formation (in the s-channel) generating a new mass scale $E_3 < E_2$. This will occur most readily in a channel in which the attraction is the most attractive. In principle the process can be repeated any number of times, leading to a hierarchy of mass scales. Such a possibility of "tumbling" has been explored in model building in particle physics. The process of tumbling, however, already exists in known phenomena. One such example is the chain: crystal formation to superconductivity, for the phonons are the Goldstone bosons resulting from crystal formation, and they in turn become the agent of Cooper pair formation in superconductors. One might even ask if the process can be continued one step further.

Another example of tumbling is found in nuclear physics [18]. First the QCD of quarks produces the mass scale of the hadrons (of the order of the so-called Λ parameter, at which the QCD interaction becomes large enough to cause chiral symmetry breaking). The nucleons and various mesons are thereby generated. The exchange of the σ meson between nucleons is attractive, and makes it possible to form nuclei out of nucleons, especially because the σ field, being a neutral scalar, can be coherently enhanced in a many-nucleon system. One might say in a nutshell that the σ is largely responsible for the existence of nuclei and for their basic properties like the shell structure, the spin-orbit interaction and the pairing, the last of which corresponds to the second stage of tumbling.

In contrast to the tumbling chain of symmetry breakings of descending energy scales, the idea of bootstrap is that the chain is circular and self-sustaining. It refers to the theoretical possibility that the Higgs field is both the cause and the effect of a BCS mechanism at the same time. The concept is similar to Chew's bootstrap hypothesis in hadron dynamics [19]. In its most general form, his bootstrap implied a duality of s- and t-channels so that the hadrons were in effect composites made out of each other. The duality principle has found its mathematical realization in the Veneziano model and the subsequent string theory.

The bootstrap BCS mechanism could in principle occur in many systems, e.g., (high T_c?) superconductors, but the specific hypothesis I have proposed concerning the $SU(2)\times U(1)$ electroweak unification means the following [20]. In the Standard Model of Salam and Weinberg, a complex doublet Higgs field is introduced to trigger a spontaneous breaking of $S(2) \times U(1)$ down to a $U(1)$ subgroup corresponding to electromagnetism. Physically realized particles are a massless gauge boson, i.e. the photon, for electromagnetism, three massive gauge bosons $W^{\pm}$ and Z^0 for the weak interactions, massive quarks and leptons (except possibly for the neutrinos), and the scalar Higgs boson. Their masses are given by

$$m_i = g_i v, \tag{4}$$

where the g_i's are appropriately defined coupling constants; $v = 246$ GeV is the vacuum expectation value (order parameter) of a component of the Higgs field, which sets the overall mass scale.

The Standard Model has so far proven remarkably accurate in describing the experimental data concerning the weak and electromagnetic interactions. No discrepancies or indications of new phenomena going beyond the model have been seen. Two input parameters, i. e., the electric charge and the Fermi constant, and a mixing angle determine v and the two gauge couplings, and thereby fix the W and Z masses, now known to be 80 GeV and 91 GeV respectively. On the other hand, the fermion and Higgs boson masses depend on Yukawa- and self-couplings of the Higgs fields which are arbitrary, so the model has no predictive power in this regard.

Recall now that, in the BCS mechanism, there was a simple 2 to 1 mass ratio between the fermion and the Higgs (σ) boson. If only one degree of freedom out of the many fermions in the Standard Model participated in this mechanism, one would expect this ratio to hold, up to renormalization corrections (which tend to reduce the ratio). It also implies that the Higgs field is a phenomenological substitute for the bound states of the particular fermion pairs in question, just as it was the case with the chiral dynamics of hadrons.

Even if this interpretation were correct, one would not know the agent that caused the Cooper pairing. Perhaps it originated in a grand unified theory which involved extra gauge fields and other degrees of freedom hidden from us at the electroweak energies. But there is another possibility, namely the bootstrap. In this case, the stronger the Yukawa coupling of a fermion, the stronger the pairing interaction, and the larger the fermion mass. Therefore one may say the heaviest one among the leptons and quarks is responsible for the formation of the Higgs field and breaking of $S(2) \times U(1)$ to $U(1)$. It is also consistent with the fact that the yet undiscovered top quark appears to be very heavy compared to all the other fermions. Current lower limits are 89 GeV for the top quark, and 40 GeV for the Higgs boson. In short, the Higgs boson is a bound state of top and antitop quarks, and their Yukawa coupling is $\gtrsim 1/3$. The Higgs boson mass should be roughly double the top mass (or less). A similar idea has been proposed also by Miransky, Yamawaki and Tanabashi [21].

The precise formulation of the bootstrap mechanism has some latitudes. Bardeen, Hill, and Lindner [22] start from a local limit of Higgs-exchange interaction between the top quark fields (as in the Fermi form of the weak interaction), and apply the BCS formalism in the standard manner, which lead to a gap equation with a quadratic divergence. The results depend on the cut-off parameter L, but in general the masses are rather large ($m_t \gtrsim 200$ GeV, and m_H is somewhat larger than m_t).

The approach I have taken [20] starts from the Standard Model as is, but treats the vacuum expectation value v as a dynamical one. Namely v is the expectation value of the potential, the so-called tadpole potential, acting on the fermion and giving it a mass due to the Higgs echange with the zero-point fermion, Higgs and the gauge fields in the vacuum. (Usually the tadpoles are regarded as a correction to a given v, and to be renormalized away, but here it is the whole contribution.) This sets up a gap equation for v since the tadpoles themselves are proportional to v, but with quadratically divergent coefficients. This is interpreted to mean that the Higgs theory is only a phenomenological representation which breaks down at the cut-off energy Λ and has to be replaced by a more fundamental theory of less divergent nature. On the other hand, the bootstrap hypothesis means that the low energy effective theory is closed and self-consistent by itself, and should not be sensitively dependent on the hidden underlying dynamics symbolized by an extra cut-off parameter. Thus one demands that the quadratic divergences cancel each other among the fermion, Higgs and gauge fields tadpole contributions. In this way, one gap equation splits into two equations, the quadratic and the remaining logarithmic part. The quadratic cancellation condition had been proposed by Veltman [22].

The two equations are constraints on v/Λ and the various coupling constants. Given v/Λ and the gauge couplings, one can determine the two unknown parameters, Yukawa coupling and the self-coupling of the Higgs field, and hence determine m_H and m_t. These equations have the form

$$\Sigma\ c_i g_i^2 Z_i^{-1} = 0 \quad , \quad \Sigma\ c_{i'} g_i^4 Z'^{-1}_i\ ln(\Lambda/m_i) = g_H^2\ . \tag{5}$$

Here the c_i's and c'_i's are numerical weights; the Z_i's and Z'_i's are renormalization constants and related quantities for the various fields contributing to the tadpoles, regarded as functions involving powers of $g_i^2 \ln(\Lambda/v)$, g_i^2 and $\ln g_i$. In the lowest approximation, one may set all the Z's equal to 1, and solve for the top and Higgs couplings g_t and g_H (or m_t and m_H) in terms of the gauge coupling's (or m_W and m_Z). One finds two sets of solutions: $m_t \sim 80$ GeV, $m_H \lesssim 60$ GeV, and $m_t \gtrsim 120$ GeV, $m_H \gtrsim 200$ GeV. Their exact values depend on Λ, getting larger for smaller Λ. The low mass solution seems incompatible with experiment. For values of Λ of the order of the Planck mass and less, the high mass solution actually gives considerably larger masses than the lower limits. However, the renormalization corrections seem appreciable, although they have not been evaluated.

It remains to be seen whether or not the basic assumptions concerning a bootstrap mechanism as the origin of symmetry breaking and mass generation in the electroweak interactions will hold up experimentally. Their tests mainly lie in the prediction of the top quark and Higgs boson masses, and the absence of early deviations from the Standard Model.

A more ambitious program would address the origin of the entire mass matrix of the fermions. From the viewpoint of the bootstrap, it is interesting that the top quark plays a special role and is by far the heaviest fermion. In fact W, Z, t and H seem to belong to the same natural mass scale of the weak interactions. So the problem is why the other fermions are so light. One has already a hierarchy problem at the current energy range. It might not be unreasonable to expect that these small masses are higher order corrections to the basic BCS mechanism proposed here. At any rate, understanding the fermion hierarchy might give one a clue to the hierarchy problem at higher energy scales.

I have benefited from informative discussions with Laurie Brown on historial literature concerning symmetry and symmetry breaking. This work was supported in part by the NSF; PHY 90-00386.

REFERENCES

[1] Y. Nambu, Prog. Theor. Phys. Suppl. **85**, 104 (1985); in *Particle Accelerators*, 1990, (Gordon and Breach, New York, 1990), Vol. 26, p. 1.

[2] L. A. Radicati, in *Symmetries in Physics (1600-1990)*, Proc. 1st International Meeting on the History of Scientific Ideas, Catalonia, ed. M. G. Doncel et al. (Univ. Autonoma de Barcelona, 1987), p. 197.

[3] For more information about the historical aspects, see, for example: Y. Nambu, Fields and Quanta 1, **33** (1970) and *The Past Decade in Particle Theory*, ed. E. C.G. Sudarshan et al. (Gordon and Breach, 1933), p. 33. *Dynamical Gauge Symmetry Breaking*, ed. E. Farhi and R. Jackiw (World Scientific, Singapore, 1982).

[4] M. Baker and S. L. Glashow, Phys. Rev. **128**, 2462 (1962).

[5] J. Goldstone, Nuovo Cimento **19**, 155 (1961).
P. Higgs, Phys. Rev. **145**, 1156 (1966).

[6] J. Bardeen, L. N. Cooper, and J. R. Schrieffer, Phys. Rev. **106**, 162 (1957).

[7] V. L. Ginsburg and L. D. Landau, Zh. Exp. Teor. Fiz. **20**, 1064 (1950).

[8] P. W. Anderson, Phys. Rev. **110**, 827 (1958); **112**, 1900 (1958).
G. Rickayzen, Phys. Rev. **111**, 817 (1958); Phys. Rev. Lett. **2**, 91 (1959).
Y. Nambu, Phys. Rev. 117, 648 (1960).

[9] H. P. Duerr, W. Heisenberg, H. Mitler, S. Schlieder and K. Yamazaki, Z. f. Naturf. **14a**, 441 (1959).

[10] Y. Nambu and G. Jona-Lasinio, Phys. Rev. **122**, 345 (1961); **124**, 246 (1961).

[11] Y. Nambu, Physica **15D**, 147 (1985); "Mass Formulas and Dynamical Symmetry Breaking", U. Chicago preprint 90-37, to appear in a festscrift in honor of S. Okubo.

[12] M. Gell-Mann and M. Levy, Nuovo Cimento **16**, 705 (1960)

[13] Y. Nambu, in *Rationale of Beings*, Festschrift in honor of G. Takeda, ed. R. Ishikawa et al. (World Scientific, Singapore, 1986), p. 3; "Supersymmetry and Quasi-Supersymmetry", U. Chicago preprint EFI 89-30, to appear in a festschrift in honor of Murray-Gell-Mann.

[14] See [11] and the references quoted therein.

[15] Y. Nambu and M. Mukerjee, Phys. Lett. **209**, 1 (1988); M. Mukerjee and Y. Nambu, Ann. Phys. **191** (1989).

[16] A. Arima and F. Iachello, Phys. Rev. Lett. **35**, 1069 (1975); Ann. Phys. **99**, 253 (1976).

[17] S. Dimopoulos, S. Raby, and L. Susskind, Nuc. Phys. **B169**, 493 (1980).

[18] Y. Nambu, in *Festi-Val*, Festschrift for Val Telegdi, ed. K. Winter (Elsevier Science Publishers B. V., 1988), p. 181; and the second reference in [11].

[19] G. F. Chew, Proc. 1960 International Conference on High Energy Physics, ed. E. C. G. Sudarshan et al. (U. Rochester/Interscience Publishers, 1960), p. 273.

[20] Y. Nambu, in *New Theories in Physics*, Proc. XI Warsaw Symposium on Elementary Particle Physics, ed. Z. A. Ajduk et al. (World Scientific, Singapore, 1989), p. 1; *1988 International Workshop on New Trends in Strong Coupling Gauge Theories*, ed. M. Bando et al. (World Scientific, Singapore, 1989), p. 2.; "New Bootstrap and the Standard Model", U. Chicago preprint EFI 90-46, to appear in a festschrift in honor of D. Dalitz; "More on Bootstrap Symmetry Breaking", U. Chicago preprint EFI 90-69, to appear in the Proceedings of the 1990 International Workshop on Strong Coupling Gauge Theories and Beyond, Nagoya.

[21] V. Miransky, M. Tanabashi, and K. Yamawaki, Mod. Phys. Lett. **A4**, 1043 (1989); Phys. Lett. **B221**, 177 (1989).

[22] W. Bardeen, C. Hill, and M. Lindner, Phys. Rev. **D41**, 1647 (1990).

[23] M. Veltman, Acta Phys. Polon. **B12**, 437 (1981).

Problems in Nuclear Physics

B. Mottelson

NORDITA, Blegdamsvej 17, DK-2100 Copenhagen Ø, Denmark

It is a special pleasure for me to be able to participate in this Centennial Symposium celebrating the pioneering contributions of Yoshio Nishina. Indeed our Institute feels a continued and living connection to the memory of Nishina, building on the ties established during his long stay in Copenhagen during the 1920's and kept alive through the efforts of his successors who have made it possible for so many of the later generations of Japanese scientists to participate in the work in Copenhagen and to so enrich the scientific life there.

I shall attempt in this report to describe some of the current activity and perspectives in the field of nuclear structure, but it seems to me appropriate on an ocassion such as this to pause for a few minutes to review the development of this field since the time when Nishina himself participated so actively and established the foundations for the subsequent development of the subject in Japan.

The study of the atomic nucleus can be seen to be divided into three rather different periods distinguished by profound differences in the nature of the questions being addressed:

I. discovery of the nucleus and its constituents (1911 - 1935)

II. defining the basic organization of the nucleus (1932 - 1952)

III. exploring the infinite richness of the nuclear many-body problem (1948 - present)

When Rutherford (1911) demonstrated that the positive charge and almost all the mass of the atom are concentrated in a particle that is extremely small compared with the size of the atom, these atomic nuclei appeared as "elementary particles" in the sense that there was no greater *a priori* reason for the existence of one of these nuclei than there was for the existence of the electron. However, as has happened so many times with other "elementary particles", it gradually became clear that the many different atomic nuclei comprise a well ordered family with relationships clearly indicating a composite structure.

The early attempts to explain the composite structure of nuclei in terms of the elementary particles then known (electron, proton, α-particle) had little success and in fact, with the discovery of quantum mechanics, was found to lead to severe difficulties. (Spin, statistics, magnetic moments, energetic confinement of electrons). The problem was, of course, that it was impossible to develop sensible ideas about the structure of nuclei until the discovery of the neutron (Chadwick 1932) and the interpretation of beta-decay by Fermi (1934) in terms of the creation of an electron and neutrino at the moment of decay.

Springer Proceedings in Physics, Volume 57
Evolutionary Trends in the Physical Sciences Eds.: M. Suzuki and R. Kubo

With Yukawa's interpretation of the force responsible for nuclear binding in terms of the exchange of massive quanta (1935), the necessary ingredients were in place, but still it required almost 20 years before the basic ground-plan of the nucleus could be delineated. I have discussed on other occasions [1] the rather violent shifts in opinions and concepts during this period and shall not attempt to repeat that discussion here. Summarizing that discussion one can see that, while the major stumbling block during the first period was lack of knowledge of the existence of the neutron, an issue in elementary particle physics, the problem in the second period was a lack of appreciation of the subtle nature of the connection between mean-free-path and elementary interactions in quantal many-body systems and of the further subtleties involved in the instability of ordered motion in the presence of small perturbations. In any case the result of the developments in the second period was the recognition that in the nucleus we have to do with a system in which the mean-free-path of nucleons is long compared with the size of the nuclei (and very long compared with the separation between nucleons) and thus a description in terms of independent particle motion is the appropriate starting point for the interpretation of nuclear structure. This conclusion is inescapable despite the striking evidence for collective, many-particle effects that had made such an impression on the scientists struggling with the issues of nuclear structure during the second period (dense spectrum of neutron resonances, occurrence of the fission process, and quadrupole moments an order of magnitude bigger than single particle size, etc.). The tension between the independent particle description and collective effects remained to be elucidated during the third period and, as I shall shortly illustrate, continues to be a recurring theme in the current studies of nuclear phenomena.

Before going on to illustrate some contemporary issues I must emphasize that the study of atomic nuclei is at the present time an activity addressing issues on a tremendously broad frontier. Besides the issues of structure and correlations such as I shall discuss, current studies of nuclei are addressing significant issues on many other frontiers, including fundamental symmetries in the strong and weak interactions, relativistic mean field effects, nuclei under extreme conditions which besides their intrinsic interest are relevant to astrophysical environments, very high energy processes in nuclei with possible relevance to a QCD deconfining transition, effects of the nuclear medium on the structure of nucleons and mesons, etc. It is clearly impossible for me to present even a very superficial survey of such a vast activity and other speakers at this symposium will address some of these issues; I shall attempt in the rest of this report to illustrate some of the aspects of current studies of nuclear structure by describing the ideas involved and the remarkable discoveries that are being made in the investigations of nuclei with very large angular momentum.

For the past 30 years, the investigation of rotational motion of nuclei has provided a rich source of information on the shapes of nuclei, on pair correlations, on the connections between collective and single particle motion, and on the fundamental dynamics of nucleons in the nuclear mean field. These studies have been greatly expanded during the past decade through the possibility of producing nuclei in states with very large angular momentum. These states are produced by allowing two heavy nuclei to collide with energies somewhat above the Coulomb barrier. In such a collision the orbital angular momentum of the relative motion may be more that 100 $\hbar$, and if the two nuclei fuse to form a compound system, this piece of nuclear matter has a correspondingly large angular momentum. The fused system is initially in a state with a very high internal excitation energy and will begin to cool by evaporation of neutrons and subsequently of γ - rays. Since these processes are much more efficient at removing energy than in removing angular momentum, the nucleus, quite rapidly, cools to the region of the lowest states compatible with the imposed total angular momentum. In these states the total excitation energy may be as much as 30 MeV (with a total level density of order 10^{21} levels/MeV) but still the states with largest angular momentum are cold since almost all of this excitation energy is tied up in generating the angular momentum; the spacing of the levels (with this high angular momentum) is similar to that in the neighborhood of the ground state and we must expect the occurrence of ordered motions, correlations and conserved quantum numbers as in the ground state region, but now modified by the controlled amount of Coriolis and centrifugal stresses imposed by the rotation. Studies of such states are exploring a wide variety of phenomena, but the scope of this report makes it necessary for me to focus on a single theme in this development - the studies aimed at identifying and resolving the quantal states associated with the very highest angular momenta in nuclei - the so-called "superdeformed states". The story falls naturally into four sections:

1. classical analysis of rotation induced deformations
2. quantal shell structure in strongly deformed and rotating nuclei
3. identification of the "superdeformed" rotational bands
4. current issues and perspectives

1 Classical analysis

The first chapter involves a classical analysis of the equilibrium shapes of rotating liquid drops [2]. This analysis mirrors the discussion on shell structure of equilibrium shapes of rotating self-attracting bodies as carried out by astronomers over the past 300 years (see Fig.1 and 2). As is well known, the result of this analysis is that for low angular momentum, I, the shape is an oblate spheroid with eccentricity proportional to I^2 (Newton, 1687). However, as discovered by Jacobi (1840), the axial symmetry is spontaneously broken at a critical angular momentum and the system develops a triaxial form developing into a bar shape before being torn apart by the centrifugal force.

2 Shell structure and deformation

For the nucleus, we know already from the study of the low angular momentum states that the classical analysis of the equilibrium shapes can be modified in a major way by the fact that the individual single particle orbits are not in general isotropic and therefore the independent particle structure will in most cases strongly favor definite non-spherical shapes that will depend on the configuration. Thus in

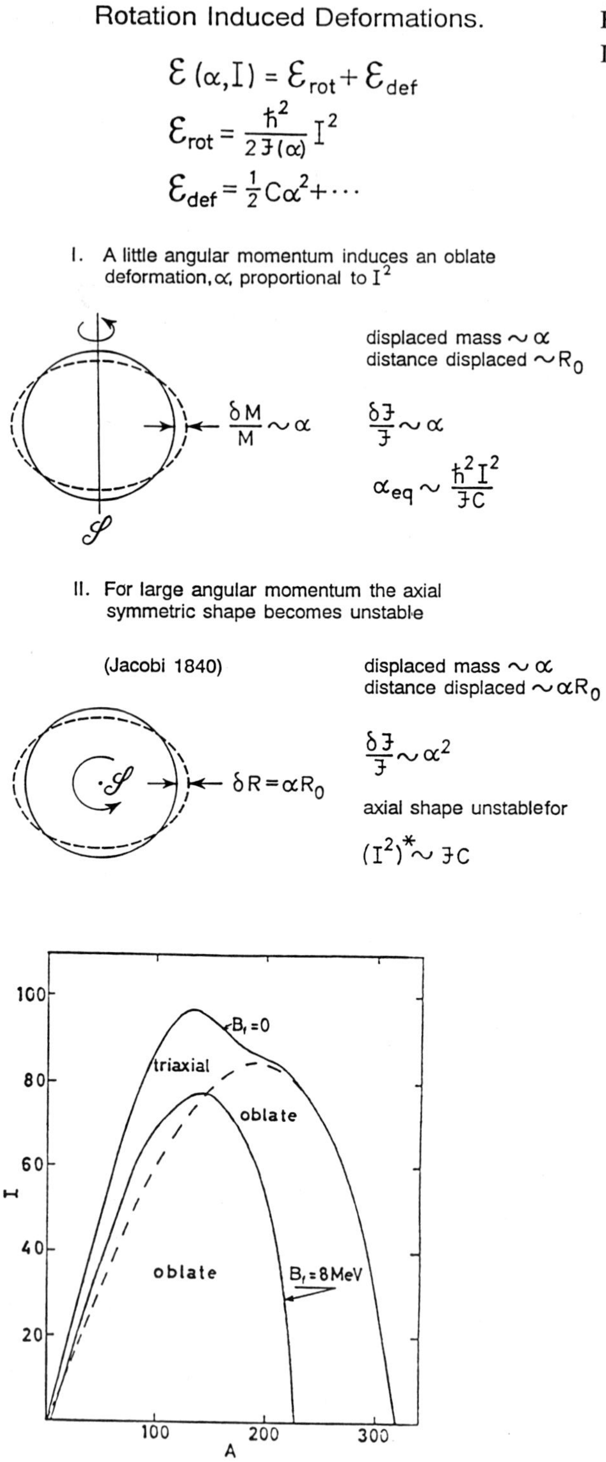

Figure 1. Rotation Induced Deformations.

discussing the expected shapes of (rotating) nuclei it is crucial to take into account the relation of shell structure and shapes.

In order to remind you of the basic concepts involved in the analysis of shell structure I would like to begin with the case of spherical systems and note that what we call shell structure is a kind of bunchiness – groups of near degeneracies – in the spectrum of single particle eigenvalues. When all the orbits in a given bunch are filled, we have a closed shell and a special stability for the spherical shape. In textbooks this bunchiness is usually discussed in a very unprincipled manner; the sequence of single particle orbits calculated in this or that potential appear accidentally and fortuitously to exhibit the near degeneracies that "explain" the particle numbers that characterize the closed shell configurations. It is clear that we should be asking more systematic questions about these matters; what is the reason for the bunching? What is the fundamental magnitude and period of these variations in the single particle level density? What should we expect if we could extend these spectra to systems with much higher quantum numbers? We must attempt to consider these more general questions if we are to discuss shell structure in very strongly deformed and rotating nuclei.

The main ideas involved in answering these questions were provided by Balian and Bloch [3] several years ago with their systematic development of an asymptotic characterization of the level density for single-particle motion in the mean field potential. This analysis focuses attention on the essential connection between the bunchiness in the single-particle spectrum and the existence of degenerate families of periodic orbits in the classical motion of a particle in the same potential.

We can see this connection perhaps most directly by considering the example of a spherical potential where the single particle eigenvalues, ϵ, depend on the quantum numbers n and l characterizing respectively the radial and angular motion. The bunchiness in the single particle spectrum occurs when it is possible to compensate the increase in ϵ associated with an increase in n by some integer amount by a corresponding integer reduction in l. This idea can be more systematically developed by considering the energy $\epsilon(n,l)$ as a continuous function of these two variables and expanding around some particular values n_o, l_o

$$\epsilon(n,l) = \epsilon(n_o l_o) + (n - n_o) + (\frac{\partial\epsilon}{\partial n})_o + (l - l_o)(\frac{\partial\epsilon}{\partial l})_o \qquad (1)$$

$$+ \text{ higher order terms.}$$

The condition for approximate degeneracy due to compensation of integer changes in radial and angular motion is

$$(\frac{\partial\epsilon}{\partial n}) : (\frac{\partial\epsilon}{\partial l}) = a : b, \qquad a \text{ and } b \text{ integer.} \qquad (2)$$

Figure 2. Stability limits for rotating nuclei. The figure is taken from [2] and shows the liquid drop estimates of the angular momentum limits for the transitions from oblate to triaxial shape (dashed line) and for the fission instability (labelled $B_f = 0$) as a function of the mass number A, selecting for each mass number the most stable charge Z.

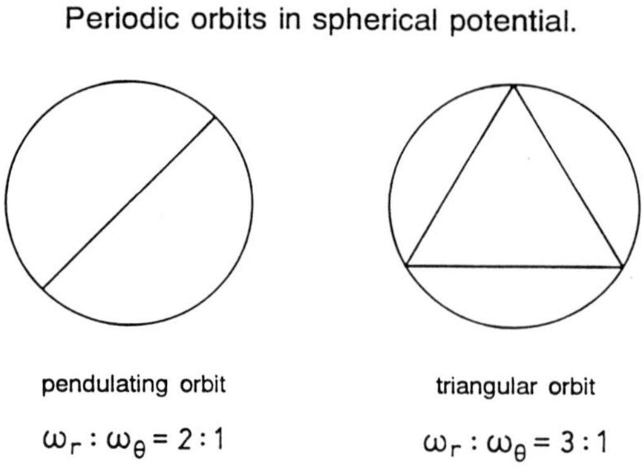

Figure 3. Periodic orbits in a spherical potential.

Under the condition (2) a change of n by b will be compensated (to leading order) by a change of l by -a units. Recognizing that the derivative of the energy with respect to action corresponds to the classical frequency we see that the condition (2) is equivalent to a condition on the classical periods of motion in the radial and angular coordinates

$$T_r : T_\theta = b : a. \tag{3}$$

When the radial and angular periods are in the ratio of integers (3), the classical orbit closes on itself after a radial and b angular periods. Thus we recognize the occurrence of the approximate degeneracies in the quantal spectrum as a reflection of the occurrence of families of closed periodic orbits in the classical motion. The degeneracies observed in the nuclear spectra reflect the two simplest classical orbits associated with periods 2 : 1 (pendulating orbits) and 3 : 1 (triangular orbits) (see Fig.3). I shall not attempt here to pursue further the very interesting consequences of this characterization of shell structure in spherical systems, but must go on with the implications of these ideas for our main theme, the shapes and configurations of the highly deformed nuclei expected for the most rapidly rotating systems. Most of what we know about this problem has been obtained from the examination of single-particle motion in the anisotropic harmonic oscillator potential (see, however, the interesting results obtained by Arvieu and Ayant [4] for motion in a spheroidal infinite square well potential). The highly simplified schematization implied by the choice of this potential should be obvious, but nonetheless it seems to provide surprisingly robust guidance for the interpretation of superdeformation and I shall therefore briefly summarize the main results of the analysis [5].[1]

In the potential

$$V = \frac{1}{2} M[\omega_z^2 z^2 + \omega_\perp^2 (x^2 + y^2)] \tag{4}$$

[1]Despite the remarkable success of these simple estimates, and the fact that there are at present no other satisfactory interpretations of the observed super-deformed shell structures, I feel that the data cry out for deeper (more generic) explanations and we may very well suspect that there lie hidden in these results some important more general principles.

all the classical orbits are periodic if the frequencies ω_z and $\omega_\perp$ are in the ratio of integers while if the ratio of frequencies is irrational there are no periodic orbits (except for the negligibly small fraction of phase space associated with motion in the equatorial plane, z $=\dot{z}$ = 0). Correspondingly the quantal spectrum

$$\epsilon(n_x n_y n_z) = (n_z + \frac{1}{2})\hbar\omega_z + (n_x + n_y + 1)\hbar\omega_\perp \tag{5}$$

exhibits very marked shell structure if and only if ω_z and $\omega_\perp$ are in the ratio of (small) integers. The case of equal frequency ($\omega_z : \omega_\perp$ =1 : 1) yields the well-known shell structure for a spherical oscillator which is built on classical orbits of elliptical form. The next simplest families are obtained for $\omega_z : \omega_\perp = 1 : 2$ for which the classical orbits vary from a figure of eight through banana shapes to a simple crescent form. The quantal spectra (5) are plotted in Fig.4 as a function of the difference between the two frequencies ω_z and $\omega_\perp$; the figure also indicates the number of particles corresponding to closed shells for the 1 : 1 and 1 : 2 potentials. The observed closed shell numbers for spherical and superdeformed nuclei are slightly displaced with respect to these harmonic oscillator predictions as a result of the more sharply defined nuclear surface which favours states of higher angular momentum and the spin orbit force which further favours among the high angular momentum states those with parallel spin and orbit.

3 Discovery of superdeformed bands

The shell structure (at 2 : 1 deformation) of Fig.4 (as slightly modified by the use of more realistic potentials), provides crucial guidance in the search for superdeformed bands but that is only a small part of the identification of these remarkable structuresemarkale; reactions of the type that I briefly described at the beginning of this discussion produce an overwhelming amount of radiation coming from more typical cascades among which are hidden the very small fraction of the transitions which carry the information concerning the level structures in which we are interested. The development of the appropriate instrumentation and experimental arrangements that make possible the separation of the wheat from the chaff has required, on the part of the experimenters, great ingenuity and sophistication to which I could hardly do justice. However, it may be useful if I briefly mention one idea that has played a significant role in the search for these patterns. In a nucleus with a highly stable shape and internal configuration the sequence of rotation states is described by the simple rigid body rotational energy expression

$$\mathcal{E}_{rot} = \frac{\hbar^2}{2\mathcal{J}} I(I+1). \tag{6}$$

The energies of the γ-transitions (connecting states differing by $\Delta I = 2$ as a consequence of the spheroidal shape of the system) are

$$E_\gamma(I) = \mathcal{E}_{rot}(I) - \mathcal{E}_{rot}(I-2) = \frac{\hbar^2}{2\mathcal{J}}.2(2I-1) \tag{7}$$

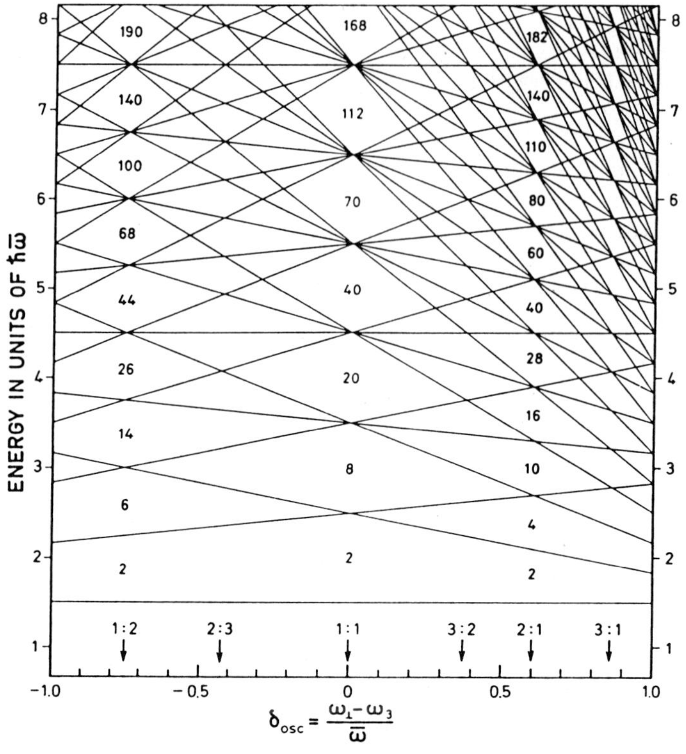

Figure 4. Single particle spectrum for anisotropic harmonic oscillator. The figure is taken from [5] and shows the spectrum as a function of anisotropy for the axially symmetric harmonic oscillator potential.

depending linearly on the angular momentum I. The differences of the γ-transition energies within such a rotational sequence

$$E_\gamma(I) - E_\gamma(I-2n) = \frac{\hbar^2}{2\mathcal{J}}.8n \tag{8}$$

are independent of I, and are integer multiples of a basic energy unit that reflects the effective moment of inertia $\mathcal{J}$ of the rotating nucleus. Thus by searching for coincident pairs of γ-rays differing in energy by integer multiples of an appropriate energy unit, it was possible for Peter Twin and his collaborators [6] to discover the spectrum shown in Fig.5. One sees here in the spectrum of $^{152}_{66}Dy_{86}$ a sequence of eighteen coincident γ-rays with remarkably constant energy differences. This sequence appears to represent the transitions between states spanning the interval from $I = 60\hbar$ to $I = 24\,\hbar$ although these numbers are uncertain by a few units since it has not yet been possible to identify the transitions connecting the "superdeformed" configurations with the (known) states with "normal" deformation at lower values of the angular momentum. That the observed sequence in Fig. 5 corresponds to a system with ratio of axes 2 : 1 is confirmed partly by the observed value of the moment of inertia $\mathcal{J}$, and even more directly by the measurement of the lifetime for

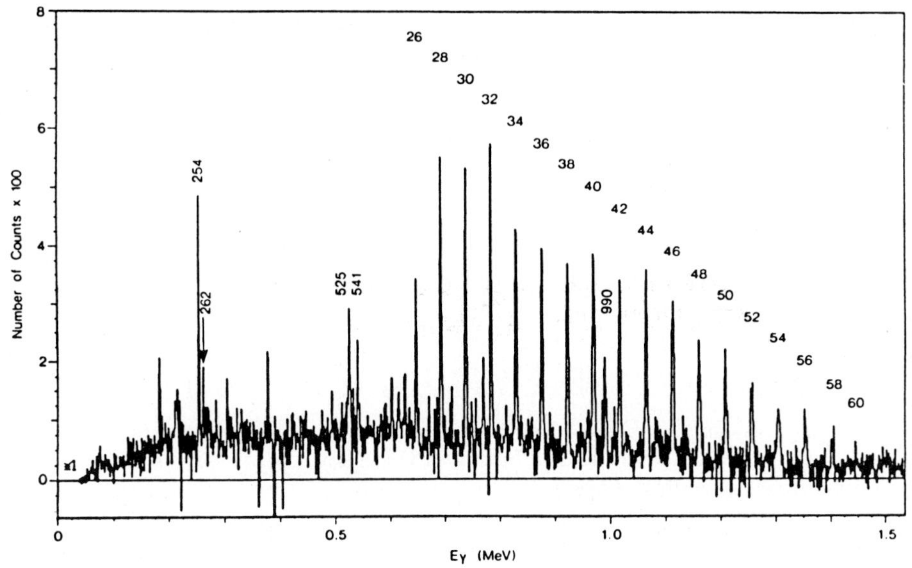

Figure 5. Gamma transitions in the superdeformed band of ^{152}Dy. The figure is taken from [6]

the emission of the observed $E2$ γ-radiation connecting the successive members of the rotational sequence.

4 Current issues

This discovery has stimulated an extensive experimental effort in laboratories all over the world; the main themes in these current investigations are:

(i) Defining the extent and regions of the superdeformed phenomena: So far about 50 examples of superdeformed bands have been identified.

(ii) Spectroscopy of the single particle configurations responsible for the different superdeformed bands.

(iii) Search for collective vibrations built on superdeformed configurations.

(iv) Characterization of the transitions connecting the superdeformed to the "normal" states of the nucleus at lower angular momentum: what is the role of superfluid effects in the tunnelling between these two phases of the nuclear system ?

(v) Observation of remarkable equalities between the γ-ray energies for transitions within superdeformed bands in neighboring nuclei: These equalities are to an accuracy of order

$$\frac{\Delta E_\gamma}{E_\gamma} \sim 10^{-3}, \qquad (9)$$

which is quite unexpected since the differences in angular momentum, mass, and radius of the involved nuclei are of relative order 10^{-2}.

(vi) Characterization of the transitions populating the superdeformed bands at the highest values of the angular momentum: what are the collective/statistical mechanisms involved in trapping of the systems in the superdeformed regions of phase space ?

These and related questions are currently providing inspiration and excitement for a very active subculture in the nuclear physics community.

References

[1] Mottelson, B.R. (1985) in The Lesson of Quantum Theory ed. de Boer, Dal, and Ulfbeck (North-Holland)

[2] Cohen, S., Plasil, F., and Swiatecki, W.J. (1974) Ann. Phys. **82**, 557

[3] Balian, R., and Bloch, C. (1971) Ann. Phys. **69**, 76

[4] Arvieu and Ayant, (1987) Jour. of Physics G **20**, 397 and 1115

[5] Bohr, A. and Mottelson, B.R. (1975) Nuclear Structure Vol. II (W.A. Benjamin)

[6] Twin, P.J., Nyako, B.M. Nelson, A.H., Simpson, J., Bentley, M.A., Crammer-Gordon, H., Forsyth, P.D., Howe, D., Mokhtar, A.R., Morrison, J.D., Sharpey-Shafer, J.F., and Sletten, G. (1986) Phys. Rev. Lett. **57**, 811

Directions of Heavy Ion Physics

P. Kienle

Gesellschaft für Schwerionenforschung (GSI),
Postfach 10 05 52, D-6100 Darmstadt 1, Fed. Rep. Germany

Abstract. Recent progress and directions of heavy ion physics in various fields of nuclear and atomic physics are presented. The progress in the acceleration technique in producing high energy high phase space density heavy ion beams in cooler storage rings is discussed. Studies of nuclei under extreme conditions address topics like the structure of nuclei at the border of nuclear stability including high spin states. Nuclear dynamics studies from the Coulomb-barrier to relativistic energies will be presented with a focus on the production of dense, heated and excited nuclear matter including the study of the properties of hadrons in such a medium particularly with respect to chiral symmetry restoration. Some atomic physics experiments with heavy ions will be addressed with emphasis on quasi-atom und e^+e^- pair production.

1. Introduction

Heavy ion research, a young branch of nuclear physics, has been exerting a growing impact on many fields of natural sciences during the past two decades. In this review we will focus on contributions to the development of nuclear and atomic physics, while interesting advances in the fields of biophysics, medicine, material science and plasmaphysics cannot be covered in the scope of this review.

The growth of heavy ion research and its scientific impact is closely related to the development of accelerators and experimental facilities for heavier ions and higher energies. In this respect two avenues have been pursued. On the one hand existing proton accelerators were upgraded to include heavy ion acceleration capabilities (e.g. Bevatron (Berkeley), AGS (Brookhaven), PS-SPS (CERN)), on the other hand many dedicated facilities have been built. With one of the latter, the UNILAC of the GSI Darmstadt, all elements up to uranium were accelerated to energies above the Coulomb-barrier already during the second half of the seventies. It was recently converted to an injector for a new built heavy ion synchrotron and a storage ring. With various cooling methods heavy ion and radioactive beams with the highest achievable phase space densities will be produced.

In the following we will first make some remarks on directions in heavy ion acceleration especially concerning recent achievements in phase

Springer Proceedings in Physics, Volume 57
Evolutionary Trends in the Physical Sciences Eds.: M. Suzuki and R. Kubo

space cooling. Then we shall focus on studies of nuclei under extreme conditions which include topics like the structure of nuclei far from stability and high spin states. Selected topics from studies of nuclear dynamics from the Coulomb-barrier to relativistic energies will be treated. We will focus on experiments to produce dense, heated nuclear matter including the study of properties of hadrons in such a medium for investigation of chiral symmetry restoration.

We will conclude with some interesting directions in atomic physics studies using heavy ions including e^+e^- pair production and various other atomic properties and processes in strong Coulomb-fields of high Z atoms.

2. Progress in Acceleration of Heavy Ions

As an introduction to the impressive scenario of available heavy ion beams Fig. 1 shows the energy charge characteristics of some typical heavy ion facilities. Linacs and cyclotrons dominate the energy regime below 100 MeV/u, whereas above this energy synchrotron facilities deliver beams up to 200 GeV/u. At the high energy frontier O^{16} and ^{28}Si beams, injected from an existing tandem facility, were accelerated in the AGS of

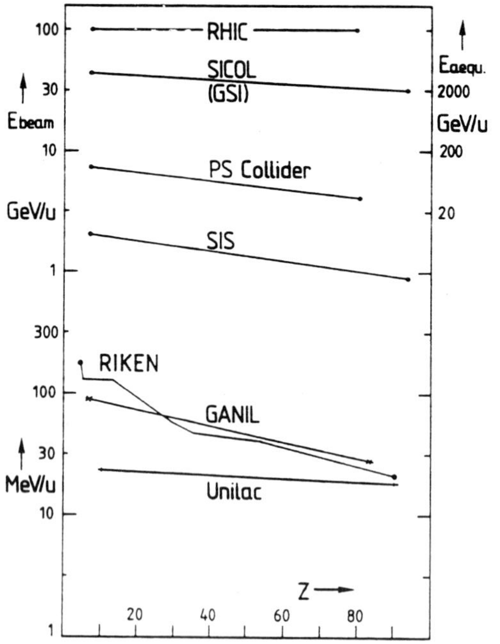

Fig. 1 Energy and ion-species characteristics of some heavy ion acceleration facilities.

Brookhaven National Laboratory to energies of 14.6 GeV/u. A booster synchrotron under construction will permit beams up to ^{197}Au to be accelerated to energies of 10 GeV/u. A GSI-LBL-CERN collaboration constructed an ion injector for the PS-SPS-CERN acceleration complex. Beams of ^{16}O and ^{32}S were accelerated to a top energy of 200 GeV/u, the highest energy beams ever produced until now. In a collaboration effort of several mainly European laboratories a new injector able to produce ion beams of heavy elements up to Pb will be constructed to become operational during 1994.

Still higher energetic nucleus-nucleus collisions may be studied in colliders. A project for the construction of the dedicated heavy ion collider RHIC is continuing in Brookhaven. A double ring structure using superconducting magnets (B_m = 3.8 T) is planned to be constructed in the 3.8 km circumference existing tunnel to reach maximum energies of 100 GeV/ per beam. The upgraded AGS will be used as injector. The most difficult problem with colliders is to achieve adequate luminosities. The luminosity design aim for RHIC is $L = 2x10^{26}$ $cm^{-2}s^{-1}$ for Au-Au collision. At CERN, the option to collide heavy ions in the proposed LHC-collider is being discussed. This would lead to collisions with 3.5 TeV/u on 3.5 TeV/u at luminosities of 10^{27} $cm^{-2}s^{-1}$. At KEK a PS heavy ion collider with beams of 4-7 GeV/u and luminosities of $10^{25}cm^{-2}$ s^{-1} is being discussed. GSI began a conceptual study of a collider with energies up to 50 GeV/u using cooled beams from an injector chain built recently to achieve as high as possible luminosities.

Fig. 2 shows the layout of the new heavy ion acceleration facility of GSI with the UNILAC as an injector into a medium energy heavy ion synchrotron (1-2 GeV/u) combined with a storage ring equipped with a powerful electron cooling device. A "cool" high current density electron beam of well defined velocity is merged with the circulating ion beam of larger spread but the same average velocity over a distance of two meters. Mott scattering provides the cooling mechanism for the ions which repeatedly traverse the continuously renewed electron beam. Fig. 3 shows the very first results [1] from cooling a 91.8 MeV/u coasting beam of Ar^{18+}-particles with a 50 keV electron beam of 1 A current in the ESR. The Schottky noise frequency spectra of the 12th harmonic of the revolution frequency of the coasting beam taken before and after cooling reveal a reduction of the momentum spread ($\Delta p/p$) from 10^{-3} to 10^{-4}. Experiments with Ar^{18+} beams of 164 MeV/u and improved alignment of the electron beam resulted in an equilibrium momentum spread of 10^{-5} after 1s of cooling. In very recent experiments a 250 MeV/u Ar^{18+} beam was cooled to a momentum spread of about $3x10^{-6}$ and a radial extension of less than 0.65 x 0.65 mm^2.

The limit of phase space density is reached when the beams make phase transitions to a condensed and finally crystalline phase as predicted by molecular dynamics calculations, some results of which are shown [2] in Fig. 4. When the plasma parameter $\Gamma = (z^2e^2/a)/kT$ (where Z^2e^2/a is equal to the Coulomb-energies of ions at the Wigner Seitz radius a and kT is the beam temperature) reaches values of about 140 the beam

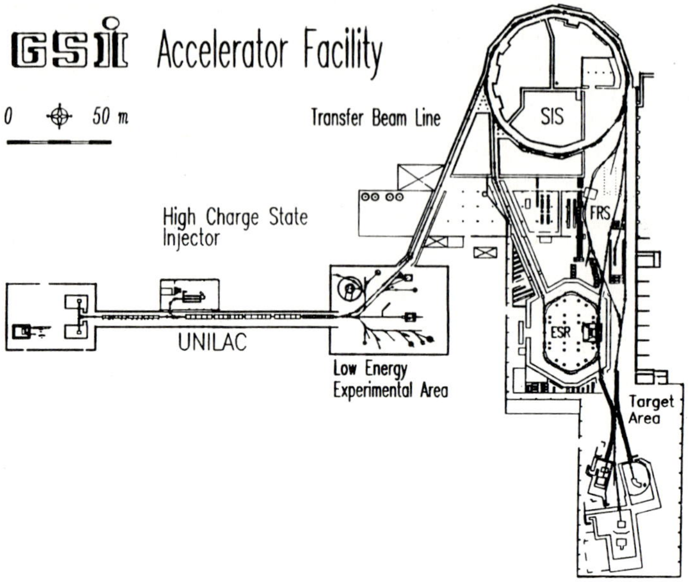

Fig. 2 General layout of the heavy-ion facility at GSI which consists of the linear accelerator UNILAC (2-20 MeV/u), the heavy ion synchrotron SIS (1-2 GeV) and the storage cooler ring ESR with half the magnetic bending power of SIS. The UNILAC can be used in parallel for a low energy program and as injector for SIS with an independent choice of energy and ion species (in 1991).

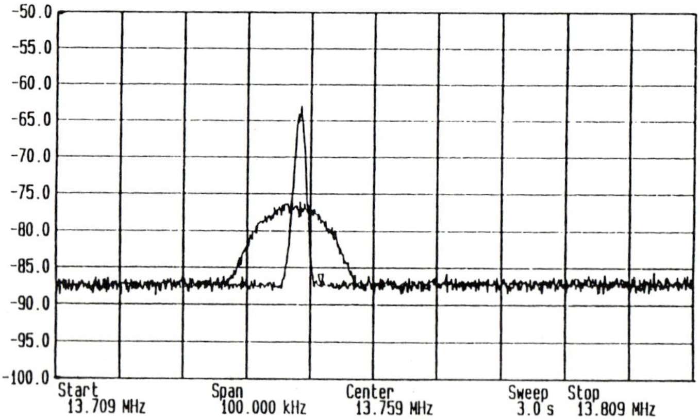

Fig. 3 Longitudinal Schottky-spectra of an uncooled (wide distribution) and a cooled (sharp peak) coasting beam at 12th harmonic of the revolution frequency are shown. Assuming the theoretical value for $\gamma_t = 2.6$, the relative momentum spread (in the logarithmic scale the full width at 3 dB below the maximum) is reduced from initially 1×10^{-3} to 1×10^{-4}. The time necessary to reach equilibrium was approximately 15 s (Ref.1)

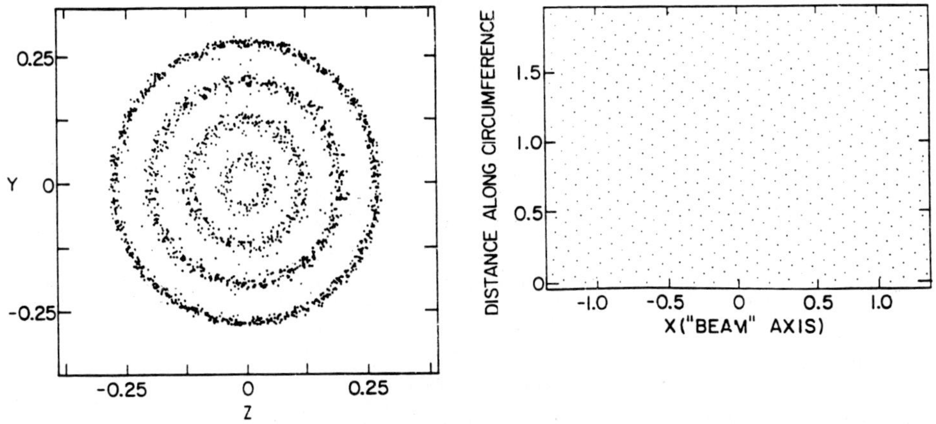

Fig. 4 Crystalline beam structure obtained in molecular dynamics calculations (Ref. 2).

particles order on concentric shells with helical structures on the shell surfaces.

Recently the first successful laser cooling of 13.3 MeV $^7Li^+$ beams was observed [3] in the Heidelberg TSR storage ring. Fig. 5 shows Schottky noise frequency spectra of uncooled (a) and cooled (b,c) beams. A revolution frequency width of 2.5×10^{-6} was achieved.

Also cooled radioactive beams of energies up to 560 MeV/u will soon be available at GSI [4] They will be produced by projectile fragmentation of SIS-beams, separated with a special fragment separator (FRS) using magnetic deflection combined with energy loss determination and will be accumulated and cooled in the ESR.

3. Nuclei under Extreme Conditions

Fig. 6 shows the chart of nuclei that are either stable (black squares) or radioactive and decay to more stable ones via β^--decay, β^+/EC-decay, α-decay or spontaneous fission. Recently several new decay modes were discovered like proton radioactivity and the emission of neutron rich light mass fragments, like ^{14}C, ^{24}Ne and ^{28}Mg. Besides the 263 stable nuclei only about 2200 of the potentially existing 6000 radioactive nuclei could be synthesized. One goal of present day heavy ion physics is to synthesize and study nuclei far from stability with unusual proton to neutron ratios up to the limits of nuclear stability against various particle emission processes like proton-, neutron-, α-decay and fission as indicated by the dashed lines.

On the neutron deficient side using heavy ion induced fusion-reactions the proton dripline ($B_p = O$) has been reached in a few cases, with the

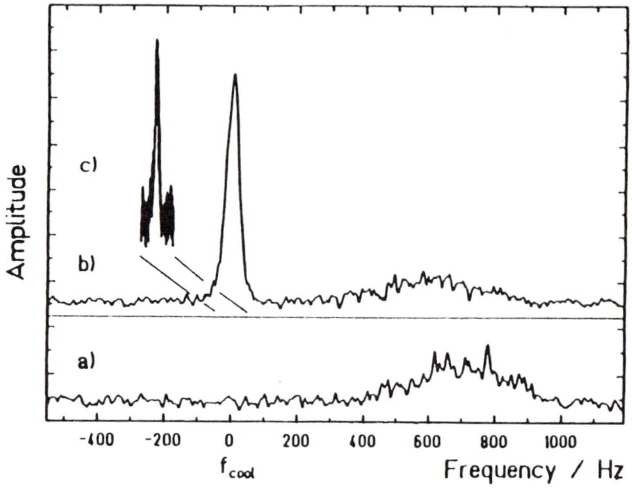

Fig. 5 Schottky noise signals. The signals are taken with a delay of 2.5 s after the injection of the ions into the TSR, the time required for the laser-cooling sweep. They are shown on a linear scale. Traces a and b are taken with identical sensitivity but different resolution bandwidths of 100 and 50 Hz, respectively, at the 18th harmonic of the revolution frequency at f_{cool} = 6.210 340 MHz. Without laser cooling spectrum, trace a is obtained, and performing the laser-cooling sweep yields spectrum b. Finally trace c, plotted with the same frequency scale, is taken at the 36th harmonic, resulting in a higher-frequency resolution (Ref. 3).

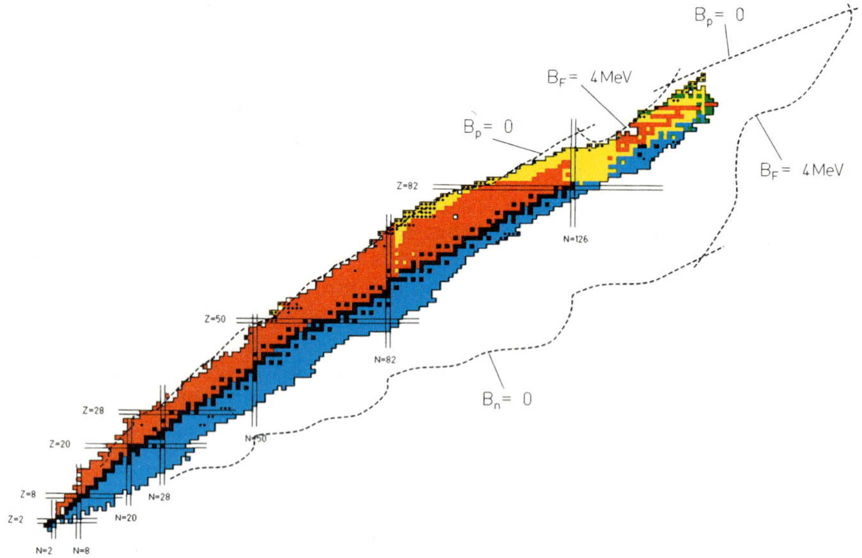

observation of proton radioactivity [5]. Future progress may be achieved using neutron deficient projectiles for the synthesis by fusion. Recently [6], a new region of deformed nuclei was discovered with the presumably double magic, N=Z nucleus ^{80}Zr in its center. Fig. 7 shows the energies of the 2^+-states of Z=N even-even nuclei from ^{64}Ge to ^{84}Mo with the corresponding ε_2 deformation parameters indicating strongly deformed ^{76}Sr and ^{80}Zr nuclei.

On the neutron rich side much progress has been made recently especially for light nuclei using projectile fragmentation in the 100 MeV energy range as synthesis reaction. It is now possible to study neutron rich light nuclei at the neutron dripline. These nuclei have very loosely bound neutrons, which may form a low density halo with properties coming close to neutron matter. It was possible to produce even beams of ^{11}Li (3 protons and 8 neutrons) with which interesting reaction studies could be performed to learn about the neutron wave functions and low lying collective excitations [7]. Fig. 8 shows the striking differences in the results

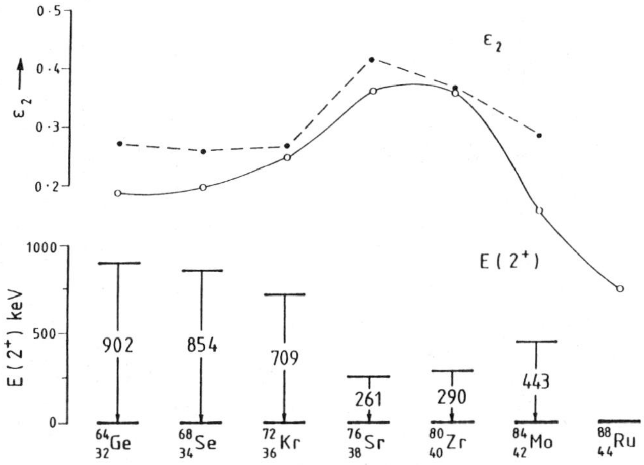

Fig. 7 The lower half of the figure shows E(2^+) for the even-even, N=Z nuclei from Ge (Z = 32) to Mo (Z = 42) (Ref. 6). In the upper half of the figure the corresponding values of ε_2, the quadrupole deformation, are shown connected by the dashed line. Theoretical predictions are connected by the solid line (Ref. 9).

Fig. 6 Island of stable or quasi-stable nuclei, defined by the dashed border contour. The black squares indicate the stable nuclei. The shaded areas contain the quasi-stable nuclei that have been produced. Indicated N and Z numbers refer to magic numbers, and doubly-magic regions are especially stable. The actinide nuclei complete the known mass table at the upper right end. The long-sought superheavy nuclei would lie around Z = 114, N = 170-180.

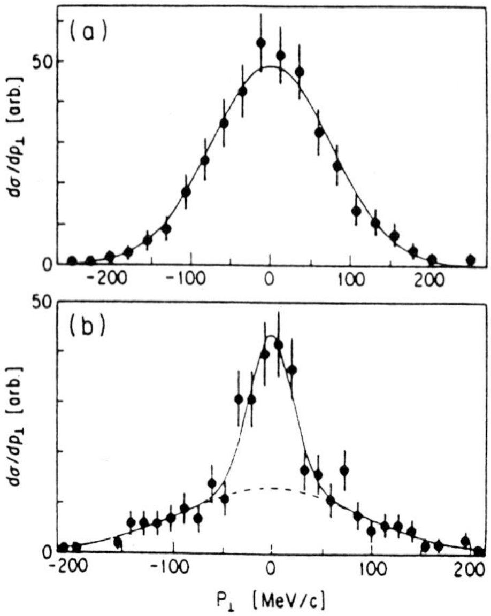

Fig. 8 Transverse-momentum distributions of (a) ^{6}He fragments from reaction ^{8}He+C and (b) ^{9}Li fragments from reaction ^{11}Li+C. The solid lines are fitted Gaussian distributions. The dotted line is a contribution of the wide component in the ^{9}Li distribution (Ref. 7).

of the transverse-momentum distributions of ^{6}He and ^{9}Li fragments from projectile fragmentation of ^{8}He and ^{11}Li nuclei respectively at 0.79 GeV/u using a ^{12}C-target [7]. The narrow component in the ^{9}Li-spectrum reflects the weak binding of the outer two neutrons in ^{11}Li. This was recently verified [8] by an experiment performed at GANIL in which the forward angular distribution of fast neutrons has been studied from fragmentation of 29 MeV/u ^{11}Li-nuclei on various targets (Fig. 9). The neutron angular distribution which is strongly forward peaked ($\Theta_{1/2} = (2.9 \pm 0.4)^{\circ}$) indicates the existence of a neutron halo with a radius of about 12 fm. For further clarification an experiment at GSI is planned in which both neutrons will be measured in coincidence with the ^{9}Li-fragments thus defining the kinematics completely. In the future more work will be done at GSI to study heavier neutron rich nuclei produced by fragmentation reactions using heavy beams. First fragmentation experiments [9] using a 760 MeV/u ^{136}Xe-beam were started to synthesize and study the double magic nucleus ^{132}Sn. One application of spectroscopic studies of properties of heavy neutron rich nuclei is related to the exploration of the path of the r-process in the nucleosynthesis of heavy elements. Here neutron rich radioactive beams may become also useful.

A further active field of heavy ion physics is the synthesis of heavy elements. The heaviest ones known until now were synthesized using cold fusion reactions bombarding ^{208}Pb and ^{209}Bi-targets with beams of ^{50}Ti,

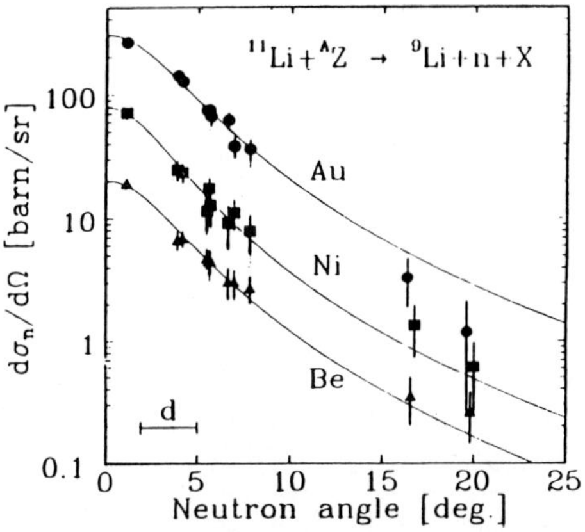

Fig. 9 Differential cross section for observation of a neutron for a ^{11}Li beam with average energy 29 MeV/u incident on Be, Ni and Au targets (Ref. 8).

^{54}Cr and ^{58}Fe at bombarding energies close to the Coulomb-barrier [10]. Fig. 10 shows the production cross sections, which drop exponentially with increasing Z. From binding energy measurements deduced from α-decay energies, fission barriers of 6-7 MeV height were discovered in these heavy elements (Fig. 11), although the liquid drop barrier has gradually disappeared. This is the outstanding discovery of this field, that shell effects lead to fission barriers sufficiently high to make these nuclei

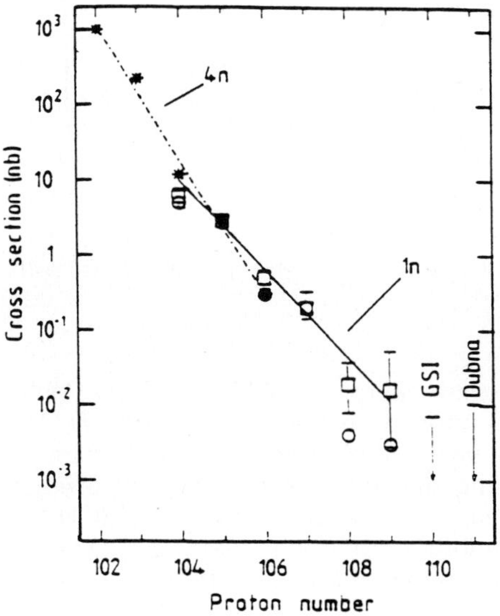

Fig. 10 Cross sections for the production of the heaviest elements. Circles, cold fusion (1n channel, Dubna); asterisks: actinide based reactions (4n channels); squares: results from GSI (Ref. 10).

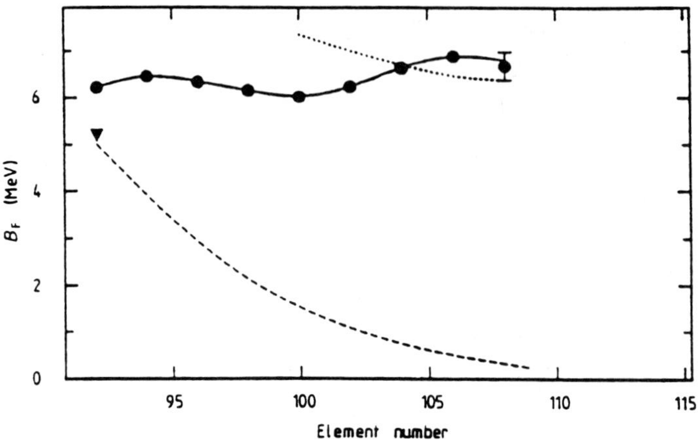

Fig. 11 Energy fission barriers for the doubly even isotopes with N-Z=48 (solid line) compared with shell corrected and liquid drop barriers.

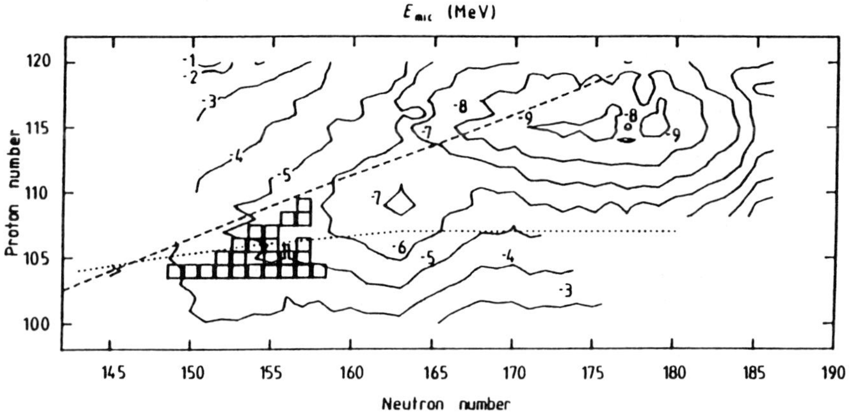

Fig. 12 Groundstate microscopic corrections of binding energies. The known isotopes of the heaviest elements are marked with squares.

relatively stable against fission; the main decay mode is α-decay. With these new mass data on high Z nuclei, one can extrapolate to still heavier elements, which are predicted to be even more stable. Calculations of shell corrections of the nuclear binding energy shown in Fig. 12 predict an island of spherical shell stabilized nuclei around Z=114 and N = 170-180. The problem is to find a suitable synthesis reaction for these nuclei. At GSI an attempt is in progress to synthesize the elements Z=110 and 111 using cold fusion reactions for which the cross sections may be extrapolated from the systematics of the data shown in Fig. 10. For the reactions of ^{62}Ni with ^{208}Pb and ^{209}Bi targets, cross sections in the pbarn range are predicted.

Besides studying nuclei at small excitation energy systematically up to the limits of stability by changing the number of protons and neutrons, heavy ion physics brought also the tools to study nuclei up to the limits of stability by adding angular momentum, by increase of their excitation energy to values up to 50-100 MeV and by raising the nuclear temperatures to values up to several MeV.

Nuclei with large angular momenta are conveniently produced in fusion reactions with heavy projectiles or in multiple Coulomb-excitations using for example ^{208}Pb-beams. In beam γ-spectroscopy introduced by Gugelot and Morinaga [11] in the beginning of the sixties made tremendous progress during the last five years with the advent of large arrays of high resolution γ-detectors.

The long predicted transition of rapidly rotating nuclei to superdeformed states were first observed [12] in 1986. Fig. 13 shows [13] the γ-ray spectrum of rapidly rotating ^{152}Dy, originating from transitions between rotational states with angular momenta up to 60 $\hbar$ and excitation energies of about 30 MeV. From the difference of the observed transition energies between states which differ by two units of angular momenta one can derive directly the moment of inertia of the rotating nucleus and its dependance on angular momentum. Together with recent measurements of the quadrupole moment of the band [13] one can infer

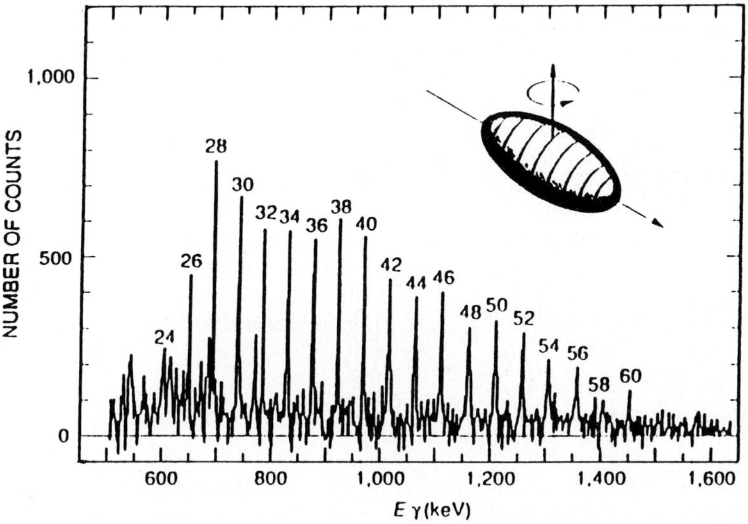

Fig. 13 Spectrum of γ rays emitted as the rapidly rotating superdeformed ^{152}Dy nucleus slows down. In such a system the γ-ray energies are proportional to the spin of the rotating nucleus, generating the regular pattern as the spin decreases in steps of two units. The spacing between the peaks yields an ellipsoidal shape having an axis ratio of 2:1. This highly deformed shape is shown in the inset (Ref. 13).

also that the shape of the nucleus is that of a "superdeformed" prolate ellipsoid of axis ration 2:1 in contrast to ground-state deformations with 1.3:1 axis ratio. Since their discovery superdeformed rotational bands were found in a variety of medium-heavy nuclei with A $\sim$ 150. Recently multiple superdeformed bands were discovered in nuclei with A $\sim$ 190. Actually the first superdeformed nuclei discovered at low spins are the so-called fission isomers in the actinide region [14], which also exhibit deformations with an axis ratio of 2:1. There are still many questions open. The population and depopulation mechanisms of the bands which start with angular momenta of 60 $\hbar$ and end at around 30 $\hbar$ are not known yet. "Hyperdeformed" bands with 3:1 axis ratios are predicted. The problem of the smooth phase transition between the strong pair correlations at slow rotation and the independent particle motion for rigid body rotation at high angular momentum is very interesting and may be attacked by studying the strength of pair transfer reactions at high spins. These and many other problems will be vigorously attacked by the coming generation of high resolution γ-detector arrays, like "Gammasphere" (USA), "Eurogam" and "Euroball" (Europe).

The spectroscopy of "hot" nuclei produced by various types of heavy ion reactions has revealed interesting facets. One is connected with the rapid loss of correlations of energies from collective transitions with increasing excitation energies signalling a phase transition from an ordered to a chaotic system. It has been known for some time from studies of proton and especially neutron resonances in nuclear reactions that the distribution of the nuclear level spacing at high excitation energies is given by radom matrix theory which describes the quantum mechanical analogs of classical chaotic systems. Much progress has been made recently, theoretically and experimentally in extending the level spacing statistics to lower energy with the result that the chaotic behaviour prevails at even low excitation.

Besides rotations, other collective excitations like giant dipole resonances were found to persist in nuclei even at elevated temperatures. Heavy ions were also successfully used to excite higher modes of giant resonances in inelastic scattering and recently at GSI experiments started to use the high frequency components of Coulomb-collisions with relativistic heavy ions to excite multipole modes of dipole giant resonances.

4. Nuclear Dynamics

Heavy ion beams with kinetic energies ranging from the Coulomb- barrier up to 200 GeV/u allow us to study various dynamic processes in nucleus-nucleus collisions which occur when nuclei interact and penetrate, ranging from slow to fast collisions with the Fermi velocity of a nucleon ($E_F \simeq$ 30 MeV) setting an useful scale. The dissipation and transport of energy, angular momentum and mass may be studied in various phenomena, which occur if one changes bombarding energy and impact parameter, the most relevant kinematic variables. In slow collisions the interacting nucleons have time to arrange themselves in the mean field of the colli-

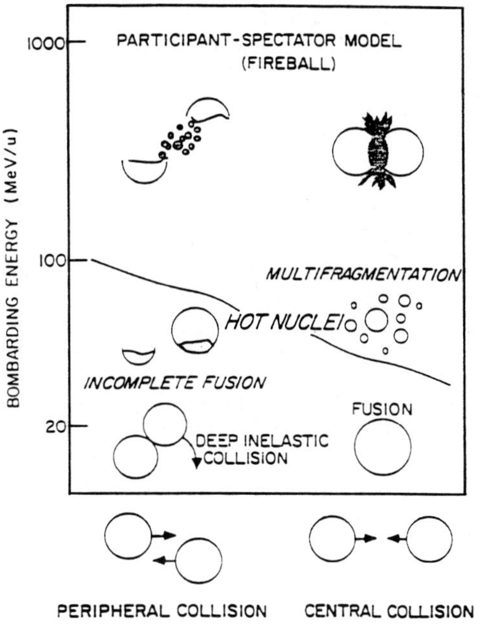

Fig. 14 Schematic representation of the different classes of nuclear reactions that can occur at different incident energies and impact parameters.

sion system, which heats up, rotates and vibrates violently. In fast collisions, nuclear matter may become compressed and heated, the nucleons may become excited and new hadrons may be created. The compressed matter expands again leading to nuclear fragmentation. The dynamics of these fast collisions are governed by the equation of state of nuclear matter at high density and temperature as well as by relaxation times and transport properties of nuclear matter under very extreme conditions. These collisions may also become a unique tool to study a fundamental property of the QCD-vacuum, namely $q\bar{q}$condensation by restoration of the chiral symmetry at high densities and/or high temperatures.

In the following we will first discuss some reaction dynamics aspects and then turn to some nuclear dynamics studies in different energy ranges.

Fig. 14 gives a schematic illustration of some of the processes which are observed and studied in nucleus-nucleus collisions in the bombarding energy range from the Coulomb-barrier to energies of about 1-2 GeV/u, for small and large impact parameters leading to central peripheral collisions respectively. At bombarding energies around the Coulomb-barrier and small impact parameters nuclei fuse, the kinetic energy and angular momentum is transfered to the compound nucleus, which may become highly excited and dexcites via the emission of light particles, photons, or especially for heavy systems by fission. In peripheral collisions, particularly of heavy systems a rotating dinuclear body is formed, in which the complete kinetic energy of the colliding nuclei is dissipated very fast on both nuclei, many nucleons may exchange before the collision partners separate again. These so-called deep inelastic collisions allow one to study dissipative transport phenomena between finite quantum systems in a unique way. At medium energy, bombarding energies and central collisions nuclei may be so highly excited that they may fragment into many pieces (multifragmentation). For peripheral collisions incomplete fusion and fragmentation may occur. At energies well above the Fermi energy a heated compressed central collision zone is expected leading to a collective flow of particles during the expansion process. At large impact parameters fragmentation is observed.

In the following we will sketch some interesting directions of nuclear dynamic studies in the low, medium and high energy regime.

At energies around the Coulomb-barrier various limitations of fusion reactions are currently being studied. At bombarding energies below the Coulomb-barrier fusion cross sections which are orders of magnitude larger than expected,using a simple potential tunnelling model (dashed line), were found in various systems [15] as shown in Fig. 15 for the complete fusion of ^{58}Ni + ^{64}Ni in the vicinity of the Coulomb- barrier. The enhancement is most likely caused by large amplitude shape vibration and transfers of nucleons by quantum tunnelling in the approach phase of both nuclei. For high Z-systems a severe limitation of the fusionability of heavy nuclei has been observed [16] with Pd-Pd being the symmetric system with the highest Z for which fusion is still observed yet with a large hindrance (Fig. 16). It is probably due to dissipation of kinetic energy in the approach phase of the nuclei, thus simulating the need for an "extra push" to overcome the barrier.

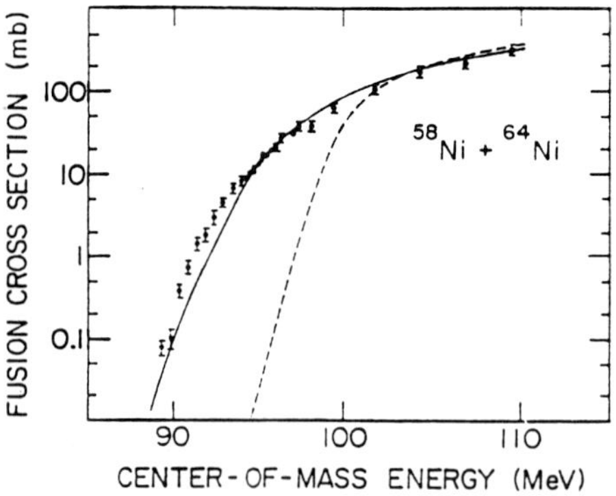

Fig. 15 Complete fusion cross section of ^{58}Ni-^{64}Ni-collisions as function of the bombarding energy. The dashed line is the cross section expected for a simple potential tunnelling model, the solid line takes account transfer and inelastic channels (Ref. 15).

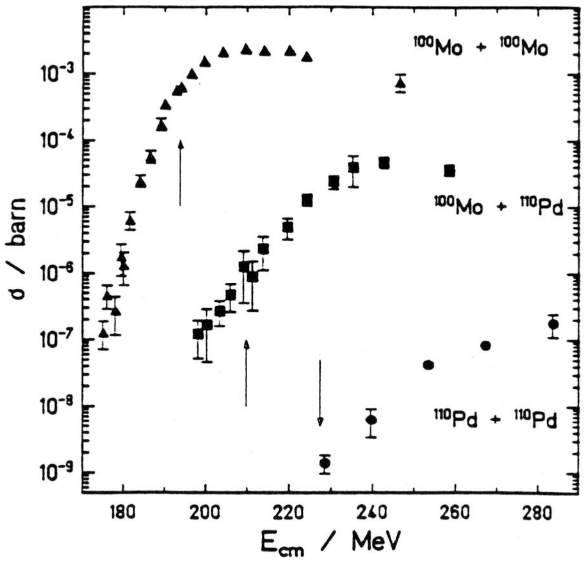

Fig. 16 Completion fusion cross sections for the heavy symmetric systems -^{100}Mo + ^{100}Mo, ^{100}Mo + ^{110}Pd and ^{110}Pd-^{110}Pd. Note the strong hindrance of fusion for the ^{110}Pd-^{110}Pd reaction (Ref. 16).

From studies of dissipative collisions between heavy nuclei various transport coefficients for energy, charge and mass-transfer between the two colliding nuclei could be determined. The interest is focussed on studying the influence of nuclear structure effects on these transport phenomena. It was found that pair correlations enhance the transfer of nucleon pairs up to an order of magnitude [17]. Coherent tunnelling of proton pairs through "weak" links of residual Coulomb-barriers of two nuclei has been predicted (Nuclear Josephson effect), but not yet observed with completely resolved states.

At medium energy, the formation of hot nuclei has been studied intensively. It is expected that if the heat supplied to the nucleus exceeds the binding energy of the nucleons, it will go into the formation of internal surfaces and the nucleus will blow up in several intermediate fragments. There were controversial experimental results on this process for a long time, but from very recent experiments at the GSI in which multifragmentation was studied in a completely exclusive experiment, a rise of the multifragmentation probability at deposited energies around 8 MeV/u seems to be indicated [18] (Fig. 17).

Central heavy ion collisions studied at the Bevalac in the bombarding energy range 200 - 2000 MeV/u with exclusive, high statistics 4π-particle detection, have shown that thermalized, compressed and heated nuclear

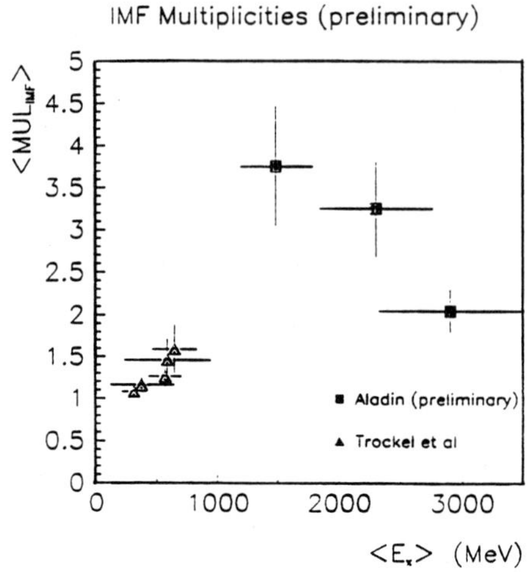

Fig. 17 Intermediate fragment multiplicities as function of excitation energies for reactions of 600 MeV/u Au ions with various targets (Ref. 18).

matter is produced. One goal is to learn about the equation of state W(ρ, T) and dynamical properties of nuclear matter from reaction observables like the rapidity distribution and especially from quantities which measure the collective flow of compressed matter in the expansion phase. Even more fundamental is the study of properties of hadrons in hot dense nuclear matter with respect to effects of chiral symmetry restoration.

Fig. 18 [19] shows the rapidity distribution of Z=1 particles from Au-Au collisions at 250, 400 and 650 MeV selected for a high multiplicity bin. For comparison a Gaussian distribution centred around midrapidity is shown. The vertical lines represent target and beam rapidity respectively. At all bombarding energies in the range between 200 and 2000 MeV/u a midrapidity fireball consisting of thermalized target and projectile nucleons with no projectile remnants is formed in central collisions. In order to investigate to which degree the kinetic energy was transformed into heat or compression a transverse momentum flow analysis of the particles was performed. Fig. 19 shows [20] the flow angle distributions of the particles from collisions of Ca+Ca, Nb+Nb and Au+Au accumulated with the plastic ball at 400 MeV/u. In the heavier collision systems Nb+Nb and Au+Au one notes for collisions with high multiplicity (central) a well expressed sidewards flow of particles similar to that expected for shockwave formation in hydrodynamical collision models. The mass and bombarding energy dependence of this collective particle flow is summarized [19] in Fig. 20. The flow is defined by the slope of the average transverse momentum per nucleon in the reaction plane as a function of rapidity determined at midrapidity, thus

$$F = \left. \frac{d < P_x/A >}{dy} \right|_{y_{cm}} .$$

Fig. 18

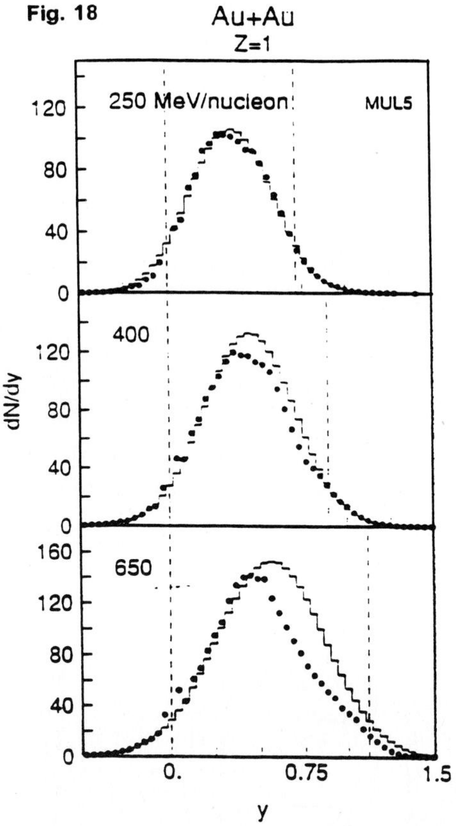

Fig. 19

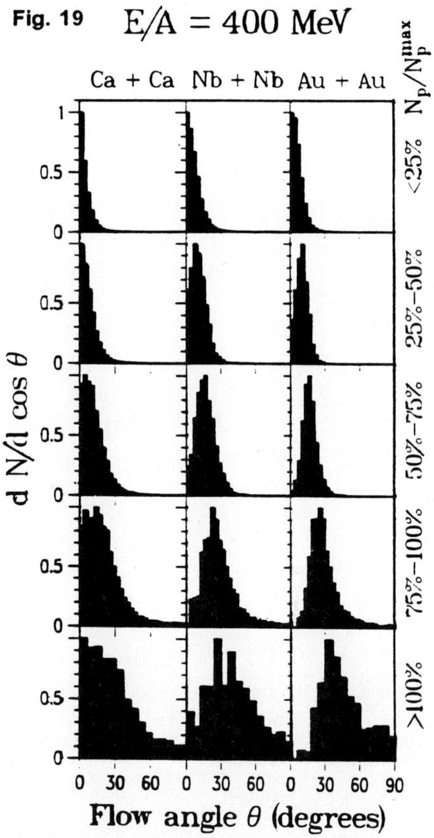

Fig. 18 Rapidity distribution for $Z=1$ particles from the reaction Au + Au at 250, 400 and 600 MeV/u. The vertical dashed lines are at target and beam rapidity (Ref. 19).

Fig. 19 Flow of particles versus flow angle Θ in Ca + Ca, Nb + Nb, and Au + Au collisions at 400 MeV/u for various multiplicities (Ref. 20).

Note that the flow rises with the mass of the system. For Au + Au and Nb + Nb-collisions a maximum around 300 MeV/u is indicated.

Out of plane flow effects have also been observed [21] with maxima also around 300 MeV/u bombarding energies. The damping of the in and out of plane flow at higher bombarding energies is interpreted as an effect caused by increased nuclear viscosity. At GSI a 4π-detector shown in Fig. 21 was constructed which allows one to measure the momentum flux of all identified charge particles in a large dynamical range. First experiments have started to make a complete momentum flow measurement for Au-Au collisions at a large bombarding energy. In addition the creation of charged and neutral mesons will be studied from far below the N-N-

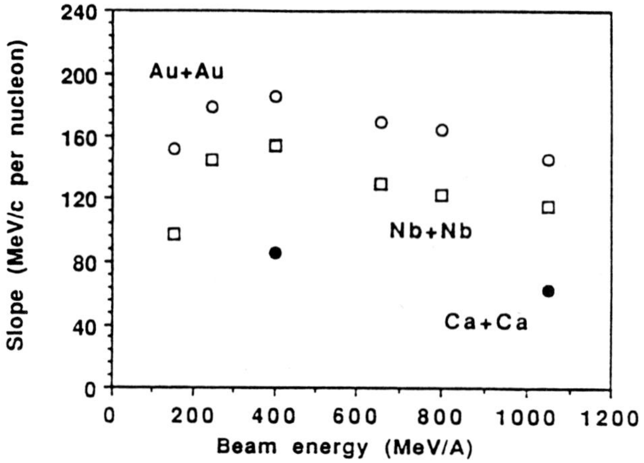

Fig. 20 Mass and bombarding energy dependence of the local slope of transverse momentum flow at midrapidity (Ref. 19).

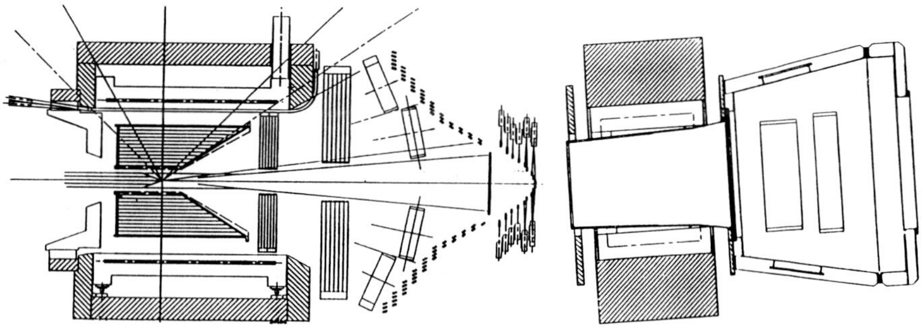

Fig. 21 Schematic layout of a 4π-detector at GSI Darmstadt. It consists of a central detector housed in large superconducting coil a forward wall with ionisation chambers and plastic scintillators and a forward spectrometer with a position sensitive multiple sampling ionisation chamber.

threshold to energies of 2 GeV/u. The neutral scalar mesons (π^O, η^O) will be detected by their two photon decays. The study of subthreshold particle production in the hot dense collision zone should also give some insight in a possible change of their effective mass and coupling constant in the hot dense medium.

The study of central collisions may be a unique tool to learn about a basic feature of QCD. The question is whether some properties of hadrons like their mass and coupling constant may be related to the order parameter of chiral symmetry restoration, which may occur in the interaction region of central heavy ion collisions. The underlying physics of this chiral

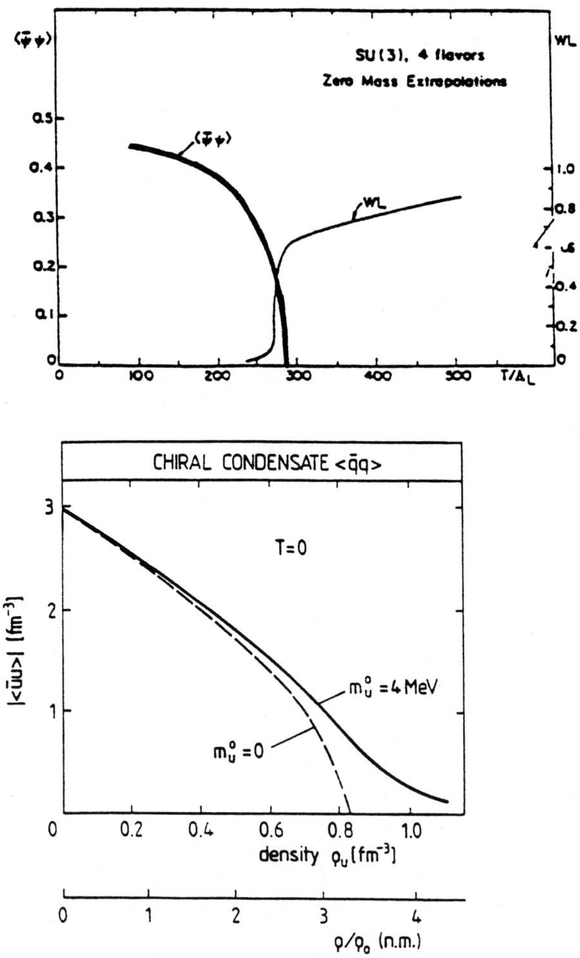

Fig. 22 Quark-condensate as function of temperature (upper part) and density (lower part).

symmetry restoration is the following. The QCD vacuum at baryon density zero and temperature zero may be thought of $q\bar{q}$ pairs like Cooper pairs in the superconducting phase because of the strong $q\bar{q}$ interaction. This causes a spontaneous breakdown of chiral symmetry and the appearance of Goldstone bosons, the pions. The order parameter characterizing this phase transition has been studied theoretically [22] as a function of temperature and baryon density. Fig. 22 shows that with increasing temperature and baryon density the order parameters of the $q\bar{q}$ condensate disappear and the chiral symmetry is restored. The question is whether one can express the hadron properties (mass, coupling constant) as function of the order parameter and by measuring these properties in dense hot nuclear media study the evolution towards chiral symmetry breaking at elevated densities and temperatures. Although there are only partial answers it may be tempting to study properties of strongly interacting hadrons by detecting their decays into photons and leptons which have only an electro-weak interaction with nuclear matter. One experiment which has been started at the Bevalac is the study of dilepton (e^+, e^-) production from the annihilation of 2 pions. By reconstruction of the invariant mass of the emitting object one may hope to study the pion mass in hot dense nuclear matter.

5. Selected Atomic Physics Phenomena

With the availability of heavy ion beams many unique atomic physics problems have been addressed ranging from the study of ionisation, electron capture, and transfer in ion-atom collisions to precision spectroscopy of transitions in highly stripped ions.

The most interesting reaction process discovered and studied at UNILAC energies is the formation of quasi-molecules and quasi-atoms in adiabatic ion-atom collisions [23]. During such a collision the electrons are exposed to a two center Coulomb-field determined by Z_1, Z_2 and the internuclear distance R(t), which is time dependent corresponding to the motion of both nuclei on Rutherford trajectories. This time-dependent Coulomb-field transfers energy and momentum to the electrons, which results in both an increase of their binding energies with decreasing R as well as an ejection of bound electrons with a certain probability. For the most strongly bound electrons with velocities close to c, there is a high probability that they will adjust their charge distribution nearly adiabatically during a slow collision ($v/c \sim 0.1$) to the two center Coulomb-field. They will thus form quasi-stationary states ("quasi-molecular" or "quasi-atomic" states) when R(t) becomes much smaller than the radius of the state considered. Of special interest are "quasi-atoms" with very high Z like Z=184 which may be formed in U-U collisions. In such atoms the binding energy of the K orbit exceeds $2mc^2$, so it becomes embedded in the negative energy continuum. The formation, characteristic binding energies and wave functions of these high Z-atoms were intensively studied by the observation of inner shell ionization and δ-ray production.

In further experiments the "ionization" of the QED vacuum was investigated by studying the e^+e^--pair production in the high time-dependent field. The surprising result was that the positron spectrum (Fig. 23) showed lines between 250 and 400 keV energy superimposed on a continuous distribution. Fig. 24 shows that the cross section for the production of the e^+ lines as well as those for continuous e^+ production rises with a high power of $(Z_1+Z_2)^{20}$, indicating a strong field effect in the production mechanism of both processes.

Measurements of e^+e^- coincidence spectra brought further surprises. They revealed narrow monoenergetic sum energy lines whereas the difference energy spectra show broader distributions (Fig. 25). A study [24] of the production cross section of the monoenergetic pairs as a function

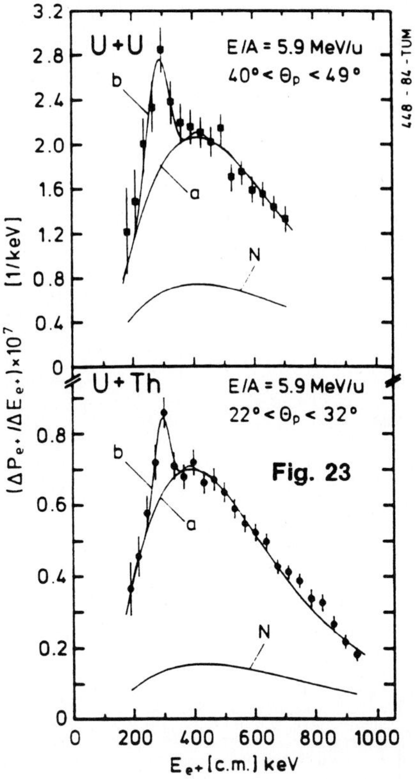

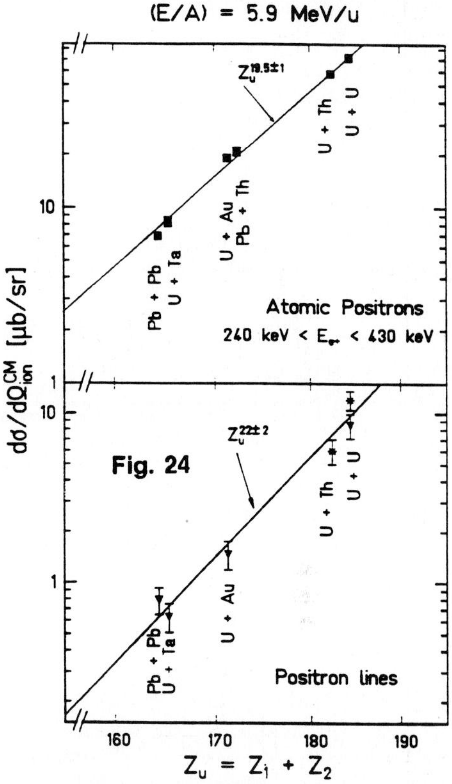

Fig. 23 Positron spectra for U-U and U-Th collisions at 5.9 MeV/u bombarding energy taken with a resolution of about 80 keV. The nuclear background N is already subtracted. Curve a represents theoretical expectations for positron creation by the strong time changing Coulomb field, with a normalisation constant 0.8 (Ref. 23)

Fig. 24 Cross section for continuous positron (upper) and positron line production as function of the combined charge $Z_U = Z_1 + Z_2$ of collisions at 5.9 MeV/u.

of the opening angle $\Theta_{e^+e^-}$ between the two leptons contained additional surprises. For the high energetic pairs a strong 180° correlation in the emission pattern was found whereas for those with lower energies all opening angles appeared. Furthermore an enhanced production probability of monoenergetic pairs for collisions in which the high Coulomb-field has a longer duration due to nuclear contact is indicated. For the moment we do not have a crisp clear explanation of all the observations. At present, our working scenario, which is under test, is about the following. In the high Coulomb-field of the dinuclear system formed in close distance collisions, composite extended objects of different masses are formed

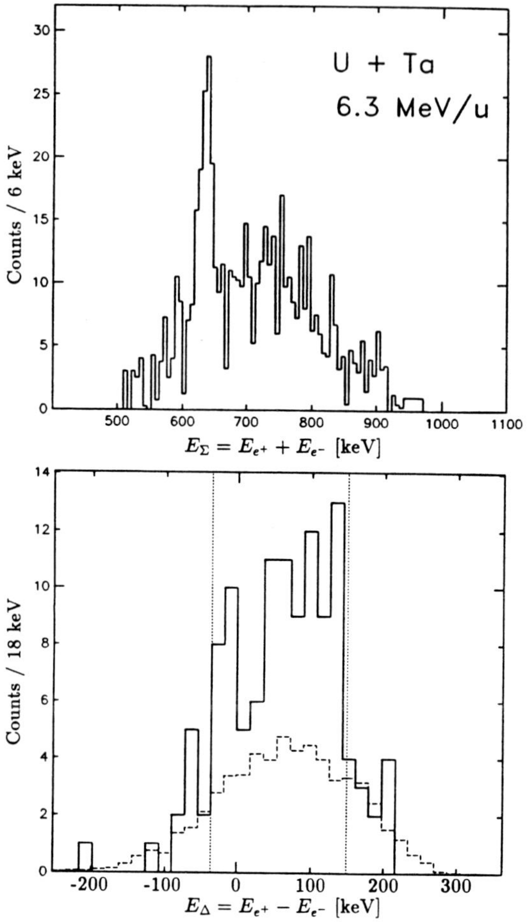

Fig. 25 Sum (upper) and difference (lower) energy spectrum of e^+e^--pairs from U-Ta collisions at 6.3 MeV/u triggered for long collision times.

which decay either unperturbed into e^+e^- pairs (180° correlations) or get dissociated by scattering on target nuclei, to which momentum is transfered but only an immeasureable amount of energy.

The heavy-ion atom collisions are also a good source for the production of highly-ionized and excited atoms, for which high precision spectroscopic data were determined (energies and lifetimes of excited states). A particular interest is focussed on measurements of the 1s and 2s Lambshift of electrons in hydrogenlike high Z atoms to determine higher order corrections of QED which come with a high power of $(Z\alpha)$ [25]. These experiments are done with increasing precision for higher and higher Z atoms.

With the availability of stored, cooled, highly ionized heavy ion beams, interesting experiments to study their interaction with laser and cold electron beams were started. Radiative capture of free electrons, spontaneous and laser induced, are studied in detail. Dielectronic recombination, a resonant electron capture process in which a free electron is scattered at a bound one, has been recently observed with high resolution and its cross section was measured.

Summary

In conclusion one notes that heavy ion physics has become a research field with a huge arsenal of accelerators and measuring devices to study very fundamental problems of QCD and QED, a great variety of structure topics of nuclei under extreme conditions and finally nuclear dynamics aspects in a wide energy span. Although the problems are complex many of them contain either fundamental aspects of interactions or of the many-body behavior of an intriguing quantum system. Thus we hope that heavy ion research has a bright future.

References

1. Franzke, B., *GSI-Report*, in press
2. Rahmann, A., Schiffer, J.P., *Phys. Rev. Lett.*, p7, 1136 (1986)
3. Schröder, S., et al., *Phys. Rev. Lett.*, G4, 2901 (1990)
4. Kienle, P., *Nuclear Matter and Heavy Ion Collisions*, ed. by Soyeur, M., Flocard, H., Tamcin, B., Poneuf, M., *Plenum Publishing Corporation 1989*, p432, and References therein
5. Faestermann, T., et al., *Phys. Lett.*, 137 B, 23 (1984)
6. Gelletly, W., et al., *Phys. Lett. B*, submitted
7. Kabayaski, T., et al., *Phys. Rev. Lett.*, 60, 2599 (1988)
8. Anne, A., et al., *Phys. Lett. B*, in press
9. Kienle, P., *Proceedings of the 8th EPS-Conference, Amsterdam* Sept. (1990), in press
10. Münzenberg, G., *Rep. Prog. Phys.*, 51, 57 (1988)
11. Morinaga, H., Gugelot, P.C., *Nucl. Phys.*, 46, 210 (1963)
12. Twin, P.J., et al., *Phys. Rev. Lett.*, 57, 811 (1986)
13. Bentley, M.A., et al., *Nucl. Phys.*, submitted
14. Specht, J.P., et al., *Phys. Lett.*, B41, 43 (1972)
15. Beckermann, M., et al., *Phys. Rev.*, C23, 1581 (1981)
16. Marawek, W., et al., *GSI-Report*, 89-1, 38 (1989)
17. von Oertzen, W., Paper presented at *Workshop on Interface between Nuclear Structure and Heavy Ion Reaction Dynamics*, University of Notre-Dame, USA, May (1990)
18. Ogilvie, C. A. et al., *Phys. Rev. Lett.*, submitted
19. Schmidt, H. R., et al., in Proceedings at the International Workshop on *Gross Properties of Nuclei and Nuclear Excitations*, IVIII, Hirschegg, Austria, January (1990), and References therein
20. Gustaffson, H.A., et al., *Phys. Rev. Lett.*, 52, 1590 (1984)

21. Gutbrod, H., et al., *Phys. Lett.*, B216, 267 (1989); Gutbrod, H., et al., *Phys. Rev. C*, in press
22. Kogut, J.B., Sinclair, D.K., *Phys. Rev. Lett.*, 60, 250 (1988); Klimpt, S., Lutz, M., Weise, W., *Phys. Lett. B*, 249, 386 (1990)
23. Kienle, P., *Ann. Rev. Nucl. Part. Sci.*, 36, 605 (1986)
24. König, W., in *Proceedings of Vacuum Structure in Intense Fields*, ed. Fried, M., *NASI-Series*, Cargése, (1990), in press
25. Gassen, J., et al., *Phys. Lett. A*, 147, 385 (1990)

The Nucleus as an Assembly of Quarks

S. Nagamiya

Department of Physics, Columbia University,
New York, NY 10027, USA

Abstract: For many years nuclear physicists treated nucleons as if they were the most fundamental constituents of the nucleus. Recently, however, the question was raised if nucleons or, in general, hadrons can retain their identity inside nuclear matter, in particular, if the matter is heated or compressed. In this talk the expected behavior of nuclear matter at very high densities and/or high temperatures is first described. Then, the current experimental efforts for the creation and detection of matter at these extreme conditions, using relativistic heavy-ion beams, are discussed and reviewed.

1. Identity of Hadrons in Nuclear Matter

Nuclei on earth have the property that their density (ρ) is almost constant: $\rho = \rho_0 = 0.17$ nucleons/fm^3 ($\simeq 0.3$ billion tons per cm^3), regardless of the species of nuclide. This means that internucleon distance (d) is almost constant; $d \simeq 1.8$ fm. On the other hand, it is known that the free nucleon has a finite radius of $(< r_{\rm N}^2 >)^{1/2} \simeq 0.8$ fm, which is only half the internucleon distance d, with its charge distribution extending even beyond its radius. This is because the nucleon is made of more fundamental particles; quarks and gluons. Therefore, once a nucleon is imbedded inside the nucleus, that nucleon might have a certain overlap with neighboring nucleons and, as a result, the wavefunction of that nucleon might be distorted and be different from that of a "free" nucleon.

This question on nucleon identity was raised first by the EMC group [1] in their deep inelastic lepton scattering experiment on nuclear targets. They found that the quark momentum distribution inside the nucleus is different from that inside the free nucleon, and hinted that the identity of the nucleon might, perhaps, be lost once it is imbedded inside nuclear matter. In my opinion, one of the most interesting questions today is the determination for those conditions for which nucleons or, in general, hadrons lose their identity so that the subnucleonic degree of freedom plays the dominant role in nuclear matter.

Springer Proceedings in Physics, Volume 57
Evolutionary Trends in the Physical Sciences Eds.: M. Suzuki and R. Kubo

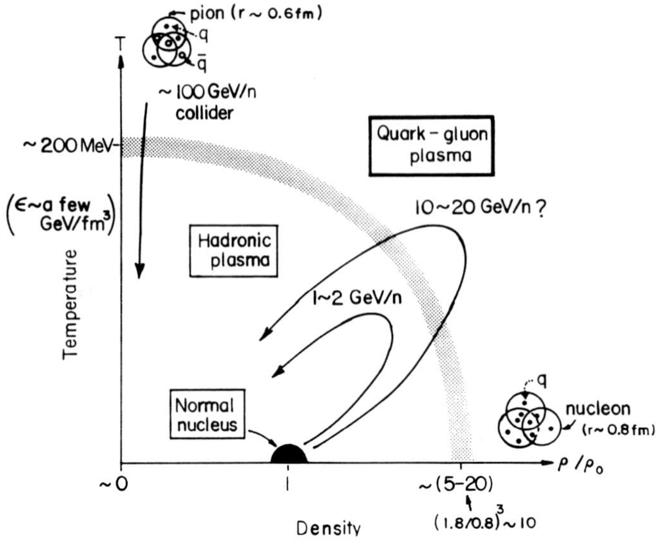

Figure 1: Expected new phase of nuclear matter and its relation to relativistic heavy-ion collisions.

If the nucleus is compressed, then the neighboring nucleons start to overlap each other more significantly. For example, at $\rho = 10\rho_0$ we have $d \simeq (< r_N^2 >)^{1/2}$: the internucleon distance is comparable to the nucleon radius. One would, therefore, expect intuitively that the nucleon would completely lose its identity at around this density and melt into a soup of quarks. Similarly, if nuclear matter is heated, then many pions are created. Since the pion is made of $q\bar{q}$ and has a finite radius of 0.6 fm, a substantial overlap among pions would be expected at sufficiently high temperatures, and the system would melt into a soup of quarks and antiquarks. The study of nuclear matter at high density and/or at high temperature is, thus, very intriguing. These soups are called quark matter, or, the quark-gluon plasma, as shown in Figure 1.

2. New Phase of Matter and Heavy-Ion Collisions

The quark-gluon plasma has two important properties. One is the Debye screening which induces deconfinement. In free space, q-$\bar{q}$ or q-q interactions are expressed by the solid curve in Figure 2. The meson or, in general, the hadron is defined as a bound state of this QCD potential. In the presence of a large number of q's and $\bar{q}$'s, however, the mutual q-$\bar{q}$ or q-q interactions are screened and weakened due to the presence

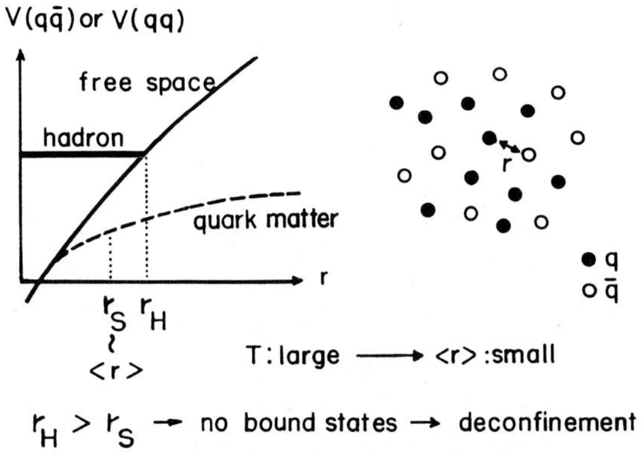

Figure 2: Mechanism of quark deconfinement in nuclear matter.

of a large number of q's and $\bar{q}$'s, analogous to the case that Coulomb interactions are screened in the presence of a large number of electrons. Now, if the temperature is increased, the number of q's and $\bar{q}$'s increases and, consequently, the screening radius decreases. At sufficiently high temperatures, if the screening radius becomes shorter than the hadron radius itself, then the q-$\bar{q}$ system can no longer be bound and, thus, is deconfined. In the deconfined phase, each quark moves almost freely inside the matter. This is the reason that the system is called a plasma.

The other intriguing property is the chirality change which induces a reduction of the effective quark mass. Most of the lattice QCD calculations [2] have shown that the order parameter increases at above the temperature of ~200 MeV to show deconfinement, and also $<\psi\bar{\psi}>$ sharply drops there to almost zero, which we call chiral symmetry restoration. More precisely speaking, a dressed quark mass changes to an undressed quark mass, which is almost zero for u and d quarks. So far, lattice QCD calculations have been done only for the zero-density high-temperature region. A similar phase transition could well be expected for the high-density region also.

Then, can we create hot and dense matter with relativistic heavy-ion collisions? Nuclear collisions at high energies can be described as two clouds of nucleons colliding with each other, by suffering many sequential nucleon-nucleon collisions. The situation is exactly like an ancient battle.

Suppose that I am a nucleon. When the battle starts, I am strongly attacked by the other party but, at the same time, I am pushed from the back by my party. Therefore, the local density around me suddenly increases. Furthermore, when I fight among others, my available energy is distributed among others, which we call thermalization in physics terminology. Therefore, the temperature of the system also increases suddenly. After the collision, everybody will be exhausted and the party disassembles. Namely, both density and temperature decrease.

In the density-temperature phase diagram, therefore, the heavy-ion collision would sweep both the high-density and high-temperature region, as shown in Figure 1. Theorists expect [3] that at a beam energy of 10-20 GeV per nucleon (hereafter, 10-20 AGeV), which is the BNL-AGS energy region, there is a possibility for the system to move into this new phase. At much higher beam energies, like 10 ATeV or 100 AGeV in a collider mode, theorists expect [3] that the colliding nuclei penetrate each other. In this case, bunched gluon strings will be created behind to form a high-temperature but zero-density region. The Relativistic Heavy-Ion Collider called RHIC at BNL, whose construction has just started this year, is regarded as an ideal accelerator to create this hot baryon-free region.

Already one may notice that one of the important experimental questions is to study at which beam energy the colliding nuclei penetrate each other, namely, they turn to grey from black. This nuclear transparency question will be discussed in the next section.

Heavy-ion beams are currently available at the BNL-AGS (15 AGeV) and at the CERN-SPS (200 AGeV) up to a mass number of $A \sim 30$. At both BNL and CERN the first heavy-ion beams were accelerated about four years ago, and much heavier-mass beams (^{197}Au or ^{208}Pb) will be available within a few years.

3. Nuclear Transparency

First, the data of energy flow together with the effort on two-boson correlations are described, both of which provide us with key information on the basic reaction dynamics. At BNL and CERN the observed charged particle multiplicity is ~200–500 [4] for $A \sim 30$ projectiles (^{28}Si or ^{32}S). Therefore, at the first stage of experimentation, all groups began with the measurement of an integrated quantity such as an energy flow.

Figure 3 shows the data from our group (E802) at BNL [5] and those from NA35 at CERN [6] on transverse energy flow, E_{T}. This energy flow is carried primarily by pions. The integrated energy flow emitted to within $90° \pm 35°$ in the nucleon-nucleon center-of-mass angles was measured in both experiments, for various targets from Al to Au.

We notice that, in the BNL energy region, the value of E_{T} starts to saturate as a function of the target mass at above Cu, whereas in the

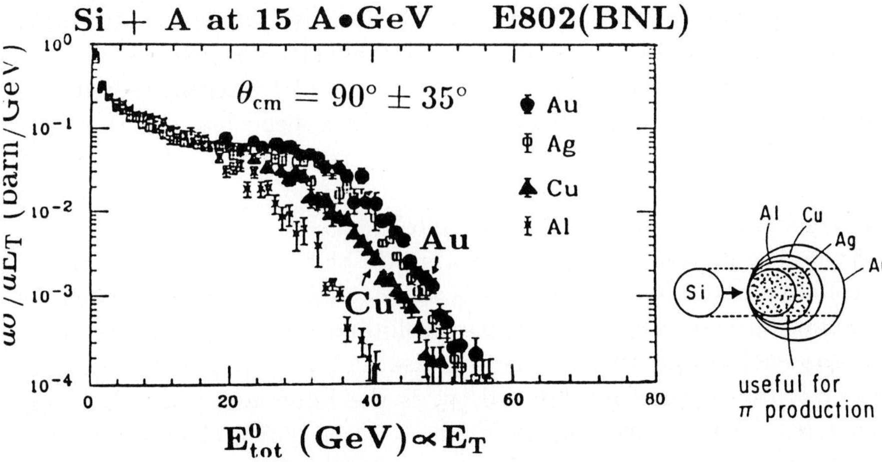

→ **Heavy nucleus seems "black" at 15 A GeV**

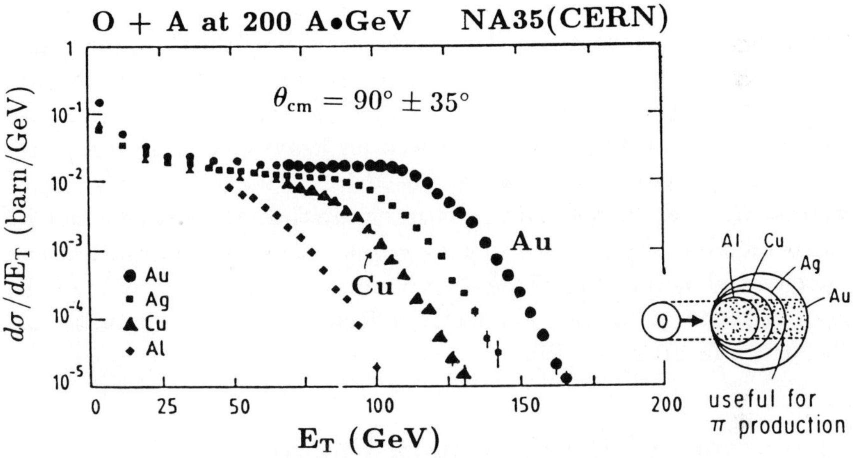

→ **Heavy nucleus seems "grey" at 200 A GeV**

Figure 3: Transverse energy flow observed at BNL (above) and CERN (below).

CERN energy domain, it increases monotonically with the target mass. The nuclear matter thickness increases by 40% from Cu to Au. The increase of E_T is, however, only 10% from Cu to Au at the BNL energy and almost nothing from Ag to Au. On the other hand, at the CERN energy, the increase of E_T from Cu to Au is ~40% which is exactly the same as the increase in nuclear matter thickness.

Naively one would expect that the thicker the nuclear matter, the more energy can be deposited into the matter, to induce a larger transverse energy flow. However, if the pion production is completed within the thickness of Cu or Ag, then the remaining matter is useless for pion production and, thereby, the saturation feature of E_T could be expected. Namely, these data suggest that a heavy nucleus seems black at 15 AGeV, whereas it seems grey or, at least partially transparent, at 200 AGeV. More detailed studies indicate that the turning point from black to grey will occur at a beam energy of ~50 AGeV.

If pion production is completed within the thickness of nuclear matter, the colliding nucleons are significantly slowed down inside the matter. For the following ideal case in which two colliding objects are slowed down and stop each other, the density will rise to $2\gamma_{cm}\rho_0$ [7], where ρ_0 is the density of the original matter and γ_{cm} is the Lorentz factor. Since $\gamma_{CM} \simeq 3$ at the BNL energy domain, there is a possibility to create nuclear matter with $\rho \simeq 6\rho_0$. The CERN energy of 200 AGeV seems slightly too high to create the high-density matter, because the nucleus seems grey, but there is still a possibility.

Then, can we probe directly the density experimentally? A long time ago an interesting method was proposed in astronomy to measure the size of a star by two-photon interferometry, called the Hanbury-Brown/Twiss method [8]. A similar method can be used for heavy-ion collisions. Here, spin-zero bosons, pions or kaons, are used for practical reasons. Once the radius, R, is determined by this method, then the density can be evaluated [9] from the relation of $\rho = m_N/V$, where m_N is the nucleon multiplicity and V $(= 4\pi R^3/3)$ is the volume which is responsible to the emission of pions or kaons. Currently, efforts to directly evaluate this density, ρ, are in progress [10].

4. Precursors of Quark-Gluon Plasma?

Now, let me describe more exotic data; one set of data from our group at BNL [11] and the other set from CERN [12].

The first one is related to the following expectation, as illustrated in Figure 4; strangeness enhancement and K^+ distillation. For simplicity, consider the zero-temperature high-density region. If the system melts into a soup of quarks, then it is made primarily by u and d quarks alone, because nucleons are made by these quarks. The Fermi energy ($E_F^{u,d}$) of this system is nearly equal to the Fermi momentum, because the masses of u and d quarks are almost zero. At $\rho \simeq 10\rho_0$ we have $E_F^{u,d} \simeq 450$ MeV. This value is larger than $2m_s c^2$ ($\simeq 400$ MeV), where m_s is the undressed s-quark mass. Therefore, it is less costly to create a $s\bar{s}$ pair than $u\bar{u}$ or $d\bar{d}$. This would induce a strangeness enhancement. In addition, if the

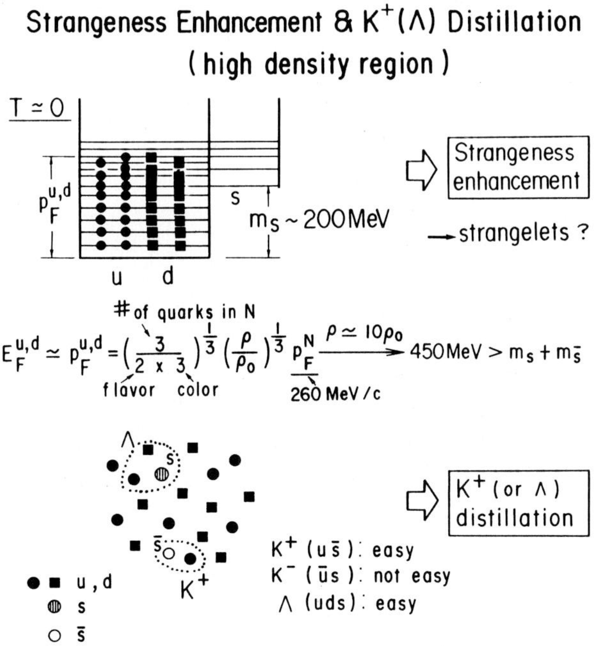

Figure 4: Proposed mechanism of strangeness enhancement and K^+ distillation for high-density quark matter.

$s\bar{s}$ is created in a soup of u and d quarks, the $\bar{s}$ can easily combine with the surrounding u to form a K^+ ($\equiv u\bar{s}$), whereas the s cannot easily find a $\bar{u}$ to form a K^- ($\equiv \bar{u}s$), because no $\bar{u}$ quarks exist in this soup. The fate of the s-quark is to combine with the surrounding u and d quarks to form a Λ ($\equiv uds$). Therefore, the K^+ and the Λ would be distilled. Since the formation of high-density matter is expected at the BNL-AGS energy domain, this prediction was, in fact, the motivation for our experiment at BNL.

We prepared a magnetic spectrometer to detect kaons. At the BNL energy domain, kaons are copiously produced in the momentum range of a few GeV/c. The identification of kaons in this momentum region is not trivial. Therefore, we spent one year developing the world's best time-of-flight counters to overcome this difficulty and, finally, we succeeded in obtaining an extremely high resolution of 75 psec, with which we were able to identify kaons up to 2.5 GeV/c.

Figure 5 shows the ratios of K^+/π^+ and K^-/π^- for three collisions, p+Be, p+Au, and Si+Au at beam energies of 14.6 AGeV. Note that pions are used for the purpose of normalization, as both π^+ and π^- yields per incident nucleon are nearly equal in these collisions [11]. The plot was made as a function of the variable called rapidity, y, which is the Lorentz

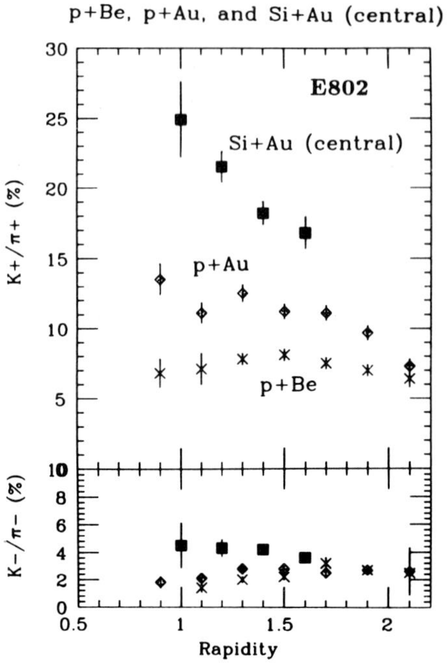

Figure 5: Observed K^+/π^+ and K^-/π^- ratios from E802 at BNL [11].

invariant longitudinal velocity. The yield of the K^- does not change too much from a very light-mass system of p+Be to a heavy-mass system of Si+Au, but the yield of the K^+ sharply increases with the mass of the system. Namely, only the K^+ is distilled by a factor of 3 or 4 from the lightest system to the heaviest system. These data are consistent with the above expectation, because, when the mass of the system increases, it is easier for the system to turn into a quark soup.

More exciting news came from CERN on the production of the J/ψ meson. Concerning this, the theoretical prediction was made by Matsui and Satz [13] before the experiment was started. The meson J/ψ is a bound state of c and $\bar{c}$ quarks. If this $c\bar{c}$ were created in the quark soup, this pair would feel the Debye screening (Figure 6). If the screening radius, r_S, is shorter than the J/ψ radius, then $c\bar{c}$ can no longer be bound to form J/ψ. Instead, the c (or $\bar{c}$) quark combines with surrounding $\bar{u}, \bar{d}$ (or u, d) quarks, when they hadronize, to form D mesons. Namely, the production of J/ψ must be suppressed, if quark-gluon plasma is created.

The NA38 group at CERN measured muon pairs to detect J/ψ [12]. Figure 7 shows the data obtained in 200 AGeV $^{16}O+U$ collisions. Clearly, the yield of J/ψ is suppressed for central collisions as compared to that for peripheral collisions. Because the quark-gluon plasma, if it exists, would

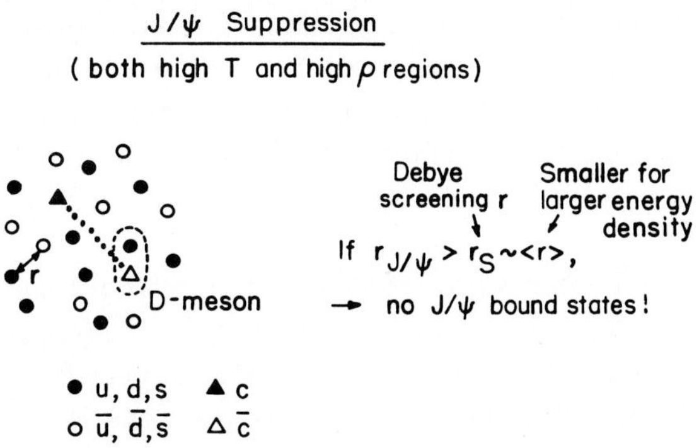

Figure 6: Proposed mechanism of J/ψ suppression for quark-gluon plasma.

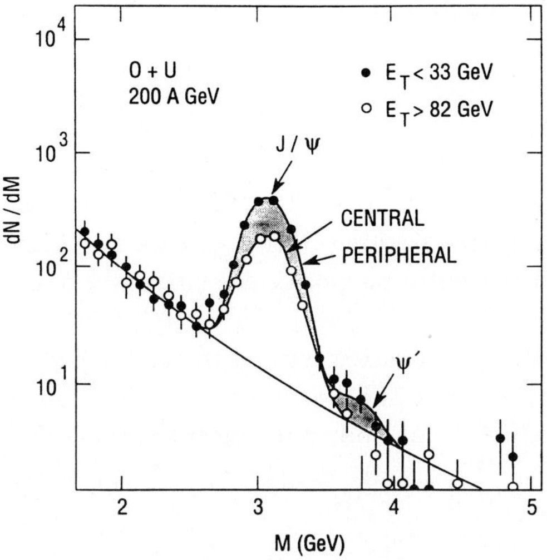

Figure 7: Observed J/ψ suppression from NA38 at CERN [12].

be created more easily in central collisions, the NA38 group suggested that, perhaps, this is the evidence for the formation of the quark-gluon plasma. By extending the measurement to ^{32}S beams [14], they further reported that the J/ψ yield is suppressed by up to 50%.

Both of these results on K^+ enhancement and J/ψ suppression are consistent with the scenario of the quark-gluon plasma formation, as mentioned above. However, mundane effects such as rescatterings inside the hadronic matter have to be carefully examined before jumping to this exotic conclusion. For example, in these collisions many pions are created and these pions may interact with target neutrons to create K^+'s ($\pi + n \rightarrow K^+ + \Lambda$). In fact, the pA data shown in Figure 5 indicate an increase in the K^+ yield as the target mass increases. Also, J/ψ absorption in hadronic matter should occur, and this absorption must be stronger in central collisions than in peripheral collisions, because a larger volume will participate in the central collision. In fact, very recent FNAL pA data [15] showed a 30% decrease of J/ψ yield in pA collisions from C to Pb target.

Many calculations taking these effects into consideration, without the formation of quark-gluon plasma, are currently in progress. Unfortunately 100% satisfactory explanations based on these mundane effects are not yet available. Clearly, a full explanation of the pA data is urgently needed.

5. Future Directions and Summary

As I mentioned before, the construction of a new accelerator, RHIC, has just started at BNL. This collider will provide, for the first time in history, a baryon-free hot region over an extended volume of a few 100 fm^3. Because the lattice QCD calculations have thus far predicted the phase transition only in the baryon-free region, RHIC will be the ideal place to study this transition. Experimental proposals are currently being actively created.

In the previous section two possible signatures related to the formation of quark-gluon plasma are discussed. For the baryon-free region, more potential signatures have been predicted. As shown in Figure 8, these are, for example, a second rise of $<p_T>$ [16], a more drastic J/ψ suppression, a low-mass lepton-pair enhancement [17], a high-p_T jet enhancement [18], an anti-matter enhancement [19], etc. Unfortunately, physics in heavy-ion collisions is not like the physics in high-energy physics where one signal pins down everything. All the predictions have certain shortcomings and uncertainties. Therefore, a coherent exhibition of these signatures as a function of a common variable is very important. If half the signatures shown in Figure 8 were observed coherently, it would already be very convincing to prove the existence of quark-gluon plasma. A detector to measure all of these has already been proposed [20].

Concerning a future possibility in Japan, I briefly mention a proposed KEK-PS Collider [21]. If heavy-ion beams are injected into the KEK-PS, then the beam energy is about 5 AGeV. In addition, if these beams are injected into a second ring, they will provide 5 AGeV + 5 AGeV in collider mode, equivalent to a fixed-target beam energy of about 50 AGeV. Therefore, it will be ideal for the study of high-density nuclear

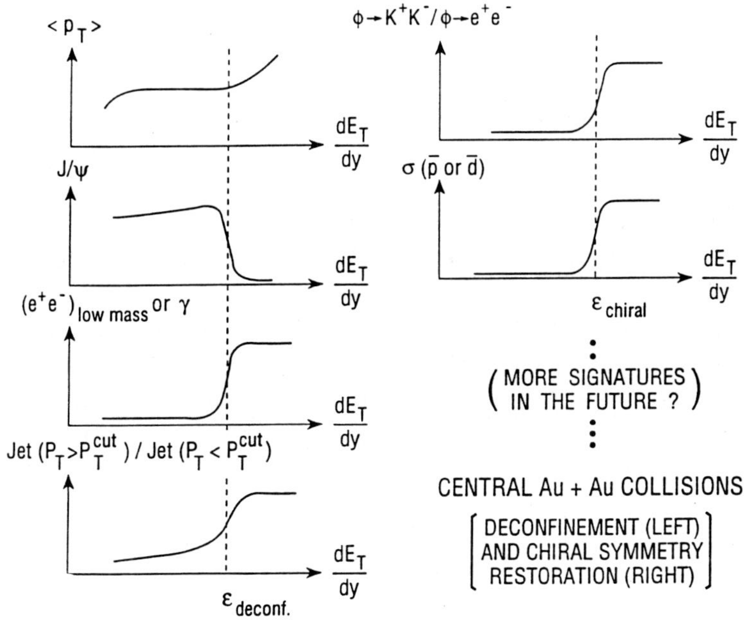

Figure 8: Potential signatures for the formation of quark-gluon plasma at the RHIC energy domain [20].

matter, being complementary to RHIC. The discussion of this PS Collider at KEK is in progress.

Let me summarize my talk. The existence of a new phase called the quark-gluon plasma is clearly predicted by lattice QCD calculations. Data on transverse energy flow show a saturation feature with target mass at the BNL-AGS energy domain but not at the CERN-SPS domain, which tells us that heavy nuclei turn from black to grey at a beam energy of about 50 AGeV. Measurements of Hanbury-Brown/Twiss correlations are in progress for both pions and kaons to directly probe the formation of high-density matter.

At the BNL-AGS energy domain, where the formation of high-density matter is expected, a significant enhancement of the K^+ meson production is observed. Also, at CERN, a 50% reduction of J/ψ production is observed. Both are consistent with the scenario of the formation of quark-gluon plasma, but more mundane effects such as the recreation of K^+ or the absorption of J/ψ might also be the reason for these anomalies. Clearly, much more careful theoretical analyses are required.

The field is young but growing. I believe that much more concrete and meaningful physics results will come out from this new field in the 1990's and into the next century.

Epilogue: In the past 100 years the concept of the nucleus has drastically changed. When Dr. Nishina was born, no scientists knew of the existence of the nucleus. In 1928, when Dr. Nishina wrote the famous paper on the Klein-Nishina formula, the existence of the nucleus was already confirmed, but it was thought that the nucleus was made of protons and electrons. When Dr. Nishina completed his first cyclotron at RIKEN in 1937, the modern concept of the nucleus was already established, and when he passed away in 1951, the basic models to describe the nucleus were about to be completed. Here, of course, the nucleon was treated as the most fundamental constituent of the nucleus. Currently, a new approach to describe the nucleus as an assembly of quarks is topical. Remarkable progress has been achieved toward understanding the nucleus during the past 100 years and, I believe, many more surprises await us in the future.

References

1. EMC Collaboration, Phys. Lett. **B123**, 275 (1983); A. Bodek et al., Phys. Rev. Lett. **50**, 1431 (1983); R. G. Arnold et al., Phys. Rev. Lett. **52**, 724 (1984).
2. For example, H. Satz, Ann. Rev. Part. Nucl. Sci. **35**, 245 (1985).
3. For example, Proc. Quark Matter '83, Nucl. Phys. **A418** (1984).
4. S. Nagamiya, Nucl. Phys. **A488**, 3c (1988).
5. M. Tannenbaum, Nucl. Phys. **A488**, 555c (1988); E802 Collaboration, Z. fur Phys. **C38**, 35 (1988), Phys. Lett. **197B**, 285 (1987).
6. NA35 Collaboration, Z. fur Phys. **C38**, 19 (1988), Phys. Lett. **184B**, 271 (1987).
7. L. D. Landau, Izv. Acad. Nauk SSSR Ser. Fiz. **17**, 51 (1953); A. S. Goldhaber, Nature **275**, 114 (1978).
8. R. Hanbury-Brown and R. Q. Twiss, Nature **178**, 1046 (1956).
9. S. Nagamiya, Nucl. Phys. **A418**, 239c (1984); Phys. Rev. Lett. **49**, 1383 (1982).
10. E802 Collaboration, preprint (1990).
11. E802 Collaboration, Phys. Rev. Lett. **64**, 849 (1990), preprint (1990).
12. NA38 Collaboration, Phys. Lett. **B220**, 471 (1989).
13. T. Matsui and H. Satz, Phys. Lett. **178B**, 416 (1986).
14. L. Kluberg, Nucl. Phys. **A488**, 613c (1988).
15. J. Moss (for E772), Talk presented at Quark Matter '90, Menton, 1990 (in press)
16. L. van Hove, Phys. Lett. **118B**, 138 (1982).
17. E. V. Shuryak, Yad. Fiz. **16**, 395 (1972); R. Hwa and K. Kajantee, Phys. Rev. **D32**, 1109 (1985); L. McLerran and T. Toimela, Phys. Rev. **D31**, 545 (1985).
18. M. Gyulassy and M. Pluemmer, Phys. Lett. **B243**, 432 (1990).
19. J. Ellis, U. Heinz, H. Kowalski, Phys. Lett. **B233**, 223 (1989); C. DeTar, preprint UUHEP-90/1 (1990).
20. OASIS Letter of Intent (S. Nagamiya, et al.) LOI #7 submitted to BNL (1990).
21. S. Nagamiya, KEK Report 88-20 (1988); Y. Mori et al., KEK Report 89-20 (1989), 90-13 (1990).

How Has Space Astrophysics Expanded the Horizon of Physics?

M. Oda

The Institute of Physical and Chemical Research (RIKEN), Hirosawa 2-1, Wako-shi, Saitama, 351-01 Japan

Abstract. Serendipitous discoveries in the 1960s drastically expanded the horizon of physics. In this paper, the evolution of major problems in astrophysics since the 1950s is analysed by decades. Several questions as of 1990 are raised. Progress of the understanding in a couple of specific subjects, i.e. the neutron star and the black hole, are briefly described. Future directions are discussed.

1. EVOLUTION OF MAJOR PROBLEMS IN ASTROPHYSICS

The decade of the 1960s was the period of so-called "Sturm und Drang" for astrophysics as represented by discoveries of quasars (QSO), 3K background of the sky, pulsars and celestial X-rays. All of the discoveries were totally unexpected, and by these, the horizon of physics has been drastically expanded. Many of the theorist's dreams or imaginary products at that time became reality. Since then nothing has become too exotic for astrophysicists and we are ready not to be surprised by anything.

In order to show how major problems in astrophysics have evolved. I would like to use a diagram for each decade, based on my own personal recollection and taste. The ordinate of the diagram towards the top represents the magnitude of surprise and to the bottom the depth of knowledge.

The abscissa is, from left to right, the universe itself, the intergalactic space, galaxies, the interstellar space and stellar objects. Such a diagram is always personal and subjective, but it hopefully is useful to help to acquire perspectives.

Springer Proceedings in Physics, Volume 57
Evolutionary Trends in the Physical Sciences Eds.: M. Suzuki and R. Kubo

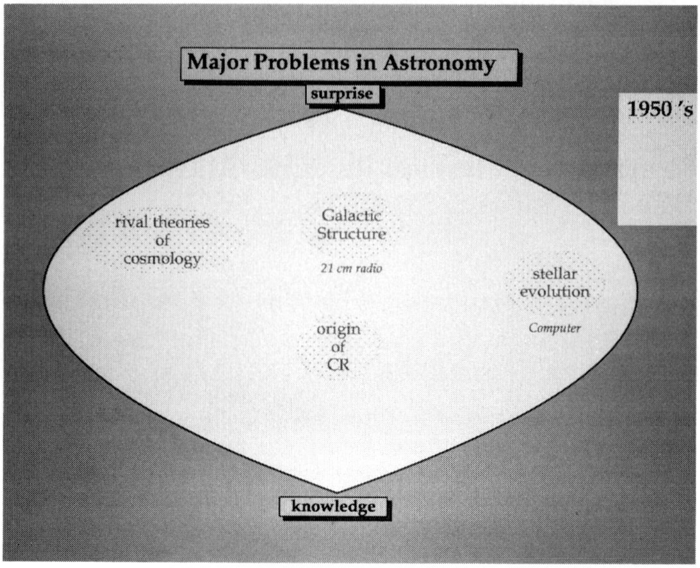

Fig. 1. Major problems in the 1950s

2.TWO DECADES FROM 1950

I recall that in the mid 1950s (Fig.1) there was a debate on cosmology between two groups, Bondi, Gold and Hoyle versus Gamow's school, i.e. the big bang school [1]. As a junior scientist I used to feel a kind of metaphysical flavor in cosmology before then, but it became real physics after that.

Also in the 1950s, radioastronomy, particularly observations with the 21 centimeter wavelength, sketched the structure of our Galaxy.

As for cosmic rays, Bruno Rossi, being inspired by Fermi, started to chase the cutoff or the maximum energy of the power law energy spectrum of the cosmic ray. In fact, during the period from the 1950s to the 1960s, for every two years of the Cosmic Ray Conference of IUPAP the upper limit increased by a factor of ten. Even now we have not yet reached the upper limit. We need a technical breakthrough to detect cosmic rays far beyond 10 ev.

The computer revolutionized the theory of stellar evolution. In Japan the Hayashi school emerged [2] and in the 1960s Icho Iben, as I remember, used up a significant portion of MIT's computer time. This was an early version of the computer experiment.

The 1960s brought many discoveries (see Fig.2). The discovery of 3K background radiation made cosmology observational physics. Quasars gave us a feeling of

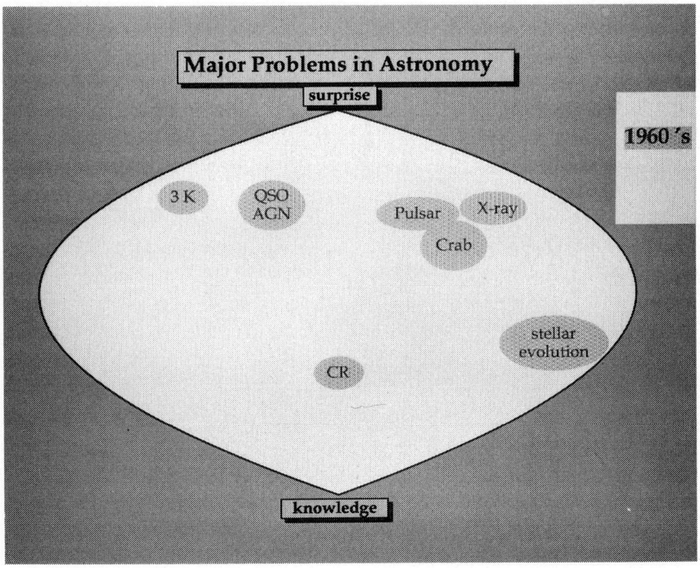

Fig. 2. Major problems in the 1960s

touching the edge of the universe, and the discovery of pulsars turned the neutron star from an imaginary to an existing star. And then the discovery of celestial X-rays by Rossi, Giacconi and others [3] made the whole picture of astrophysics different. I would like to quote one sentence from Freeman Dyson's essay [4]. It reads "The old quiescent universe of Aristotle, which had survived essentially intact the intellectual revolutions associated with the names of Copernicus, Newton and Einstein, disappeared forever as soon as the X-ray telescopes went to work".

As for the Crab Nebula, Woltjer [5] regarded it as nature's laboratory for physics and since then it has continued feeding us physics, being particularly stimulated by the new emergence of X-ray astronomy.

3.THE 1970s AND THE 1980s

In the 1970s (Fig.3), X-ray astronomy opened new chapters in astrophysics. First of all, using the observation of the X-ray stars or compact X-ray objects, we learned a lot about the neutron stars. Now the neutron star is not an imaginary subject. Also a strong candidate for a black hole, Cyg X-1, emerged. The cosmic X-ray background which had been observed since the early period of X-ray astronomy had been studied more deeply.

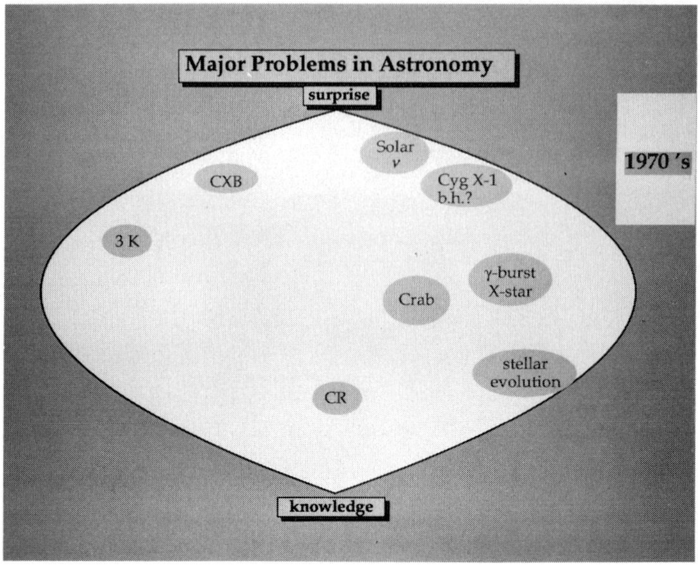

Fig. 3. Major problems in the 1970s

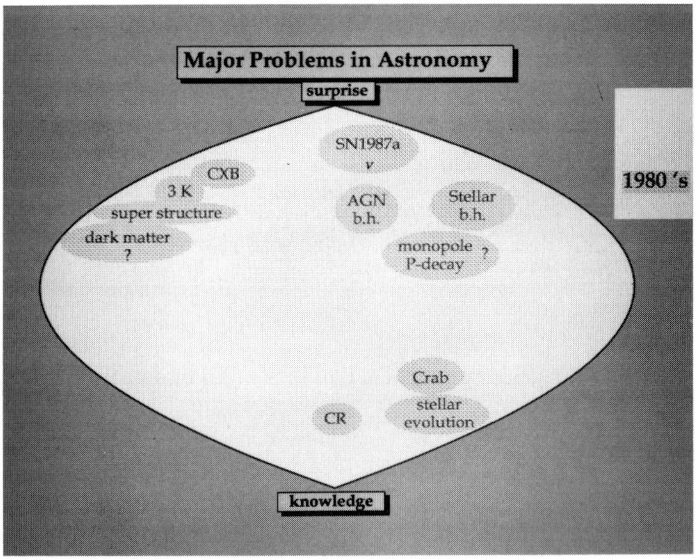

Fig. 4. Major problems in the 1980s

In the 1980s (Fig.4), by eliminating various possibilities, active galactic nuclei or centers of QSO became very good candidates for so-called giant black holes.

The famous experiment of Davis to detect solar neutrinos was initiated in 1970 [6].
In the late 1980s, of course, the supernova SN 1987a was a real surprise, though from the statistics of one supernova per several centuries we had jokingly expected a supernova during our lifetime. This supernova is continuing to give us many projects and we could have produced many young PhDs from this.

Figure 5 is a cartoon version of a Christmas card of 1987 in which early observations of SN 1987a are summarized. Immediately after the bang KAMIOKANDE-II observed eleven neutrinos within a few seconds [7]. The Irvine-Michigan-Brookhaven people had another big tank underground. A Japanese X-ray astronomy satellite which was launched only three weeks before the supernova observed the appearance of X-rays, about half a year after the bang while many theorists predicted that X-rays should appear one year or two years later [8]. A collaborative group of Russians and Germans have observed X-ray as well [9].

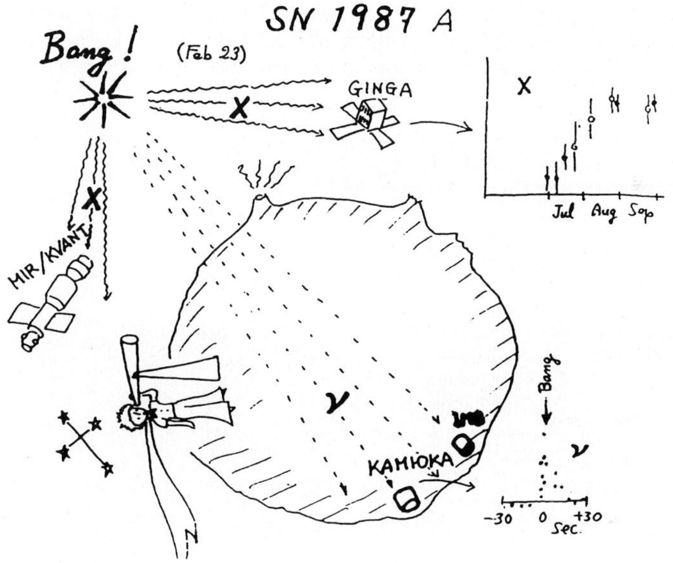

Fig. 5. A cartoon which summarizes immediate observations after SN 1987a

4. REMAINING MAJOR QUESTIONS

As of now, there remain many unsolved questions (see Fig.6). No monopoles and no proton decay have been found in contradiction to expectations produced by the Grand-Unified-Theory regardless of various experimental efforts. The origin of cosmic rays or the upper limit of the cosmic ray energy has not yet been identified. There have been many theoretical proposals for the dark matter, but they all have disappeared or been proven to be unsuccessful.

The latest and most influential observation is that of COBE. Not only cosmology but many other theories regarding the universe [10] or the edge of the universe have to be accommodated to the COBE's results.

The original Davis experiment of solar neutrinos of 1970 and the KAMIOKANDE II observation agree with each other in one aspect but not in the aspect of the relationship with the solar cycle. We hear with surprise the latest news that the preliminary observations of the Soviet-American experiment at Baksan, if correct, did not confirm these two previous results [11].

As for the cosmic X-ray background, the GINGA satellite, which is now actively working, confirmed that the superposition of the dark active galactic nuclei is close to one third or one half or maybe even one hundred percent of the cosmic X-ray background [12]. From the energy spectrum of the 30-60KeV range

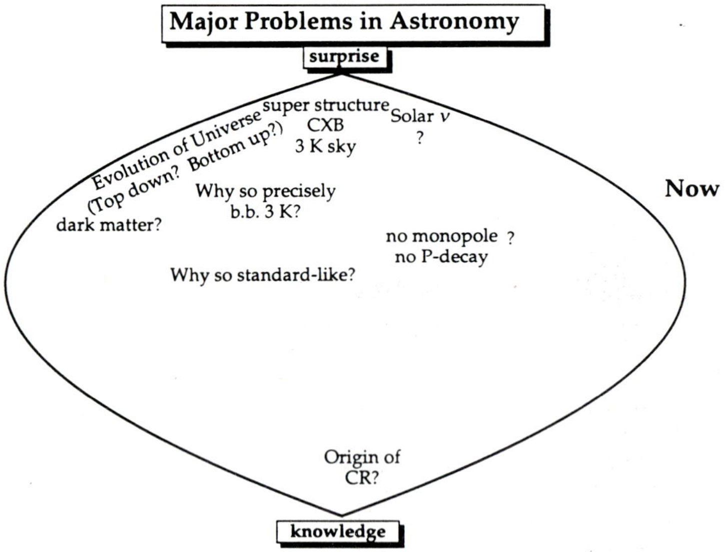

Fig. 6. Major Questions now

of cosmic X-ray background we know that the thermal bremsstrahlung of 40eV electrons in the hot plasma beautifully explains the spectrum [13]. But, if so, the total amount of hot gas is very close to the critical density. And then, due to the Compton scattering the spectrum of the microwave background in the whole sky should be distorted. But COBE forbids the distortion. Here is a clear contradiction.

5.NEUTRON STARS AND BLACK HOLES

Two specific subjects will be picked up in the following. One is the neutron star. It was advocated as a hypothetical or imaginary object by, say, theoretical prophets starting from Baade-Zwicky, Oppenheimer, and others, and then believers have followed them.

Since the mid 70's, its existence has no longer been questioned and many users appear to regard the neutron star as a laboratory for highly condensed matter physics.

We clearly know by observations that the mass of the X-ray pulsar is between one to two solar masses. And also we know that the radius of the X-ray burster is 10 km. These figures fit theoretical predictions for the neutron star surprisingly well [14]. Absorption lines in the spectrum of some X-ray pulsars, assuming that they are cyclotron lines, lead to the magnetic field strength of 10^{12} gauss.

There is another exotic object: gamma ray burst sources, the spectra being extended to a range of higher energy. We see them some ten times per year and they last about 10 to 20 seconds. There is no optical or radio counterpart. The gamma ray burst is still a kind of mystery. Recently GINGA clearly showed the cyclotron line absorption in the gamma ray burst which led to a magnetic field equal to that of the neutron star [15].

Many observers study the interaction between a neutron star and its environment under extreme conditions of magnetic field, temperature, gravity and matter density. The Crab Nebula again is a very good laboratory [16] and we have some hope that the spinning neutron star of the Crab Nebula may explain the acceleration of the particles to 10^{14} eV.

The second specific subject is the black hole. The black hole has been an even more exotic subject than the neutron star. Of course by definition it is hard to observe. But now, if a vote were taken, few astrophysicists would doubt its existence [17]. GINGA is producing a list of candidate black holes [18]. Sometimes we even drop the word "candidate". Now, is there any fingerprint for a black hole? We know that it shows a very wild flickering and different energy spectrum from other X-ray sources.

If one looks at the time sequence of the intensity of the black hole candidates, one may see that it is an example of a fractal and this might reflect the space-time structure near the event horizon or ergosphere around the black hole. It is not easy to characterize apparently aperiodic time variation of X-ray sources with, for example, power spectrum analysis. The question is if one can detect any regularity in an apparently white noise.

One of my former colleagues, Dr.K.Mitsuda, indicated that, if the time sequence of the X-ray intensity is recorded with a music tape-recorder and one listens to the "music", one source may be distinguished from another more sensitively than by the analytical method: e.g. one source may sound more metallic than the other. Of course, the method is premature and subjective, but I find it interesting. I would like you to listen to the tape. (Small pieces of the music are exhibited in this text by the sonagram in Fig.7, which is not at all as impressive as the "music".)
Current and expected observatories or observations in the near future are indicated in Fig.8. We may categorize various types of approaches which are equally important. One is, of course, the orthodox approach like big telescopes, the Hubble Space Telescope, and AXAF. Next is the adventurous approach like COBE, HIPPARCOS, gravitational wave detectors and others. The third is, using the terminology of Freeman Dyson, the "small but quick is beautiful" [4] approach,

Fig. 7. Sonagram of the X-ray star music

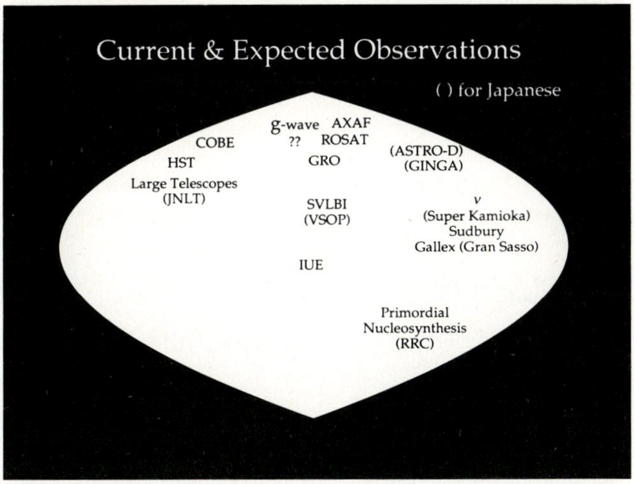

Fig. 8. Current and planned observations

in which emphasis is placed on the continuity of the programme. We should remember in any case that very often important discoveries in astrophysics have been serendipitous.

In order to show an example of the evolution of disciplines, a chronological chart is exhibited in Fig.9 to represent the history of X-ray astronomy since 1963. In the chart space vehicles are also indicated. As an example of the "small but quick is beautiful" approach, a few photographs are exhibited in Fig.10 to show the launching of the X-ray astronomy satellite "GINGA" [19]. GINGA, which stands for the galaxies, was produced in a collaboration among ISAS of Japan, Leicester University, MSSL and RAL of U.K. and LASL of the U.S. and was launched in 1987 three weeks before the supernova SN 1987a and is currently in operation in international collaboration. The photographs show students, young professors and technicians of ISAS at work.

Acknowledgements

I would like to thank Profs.Y.Tanaka, F.Makino, F.Nagase and H.Inoue of the Institute of Space and Astronautical Science and K.Makishima of the University of Tokyo for providing valuable materials and discussions for this paper and to Dr.N.Kawai for preparing the diagrams.

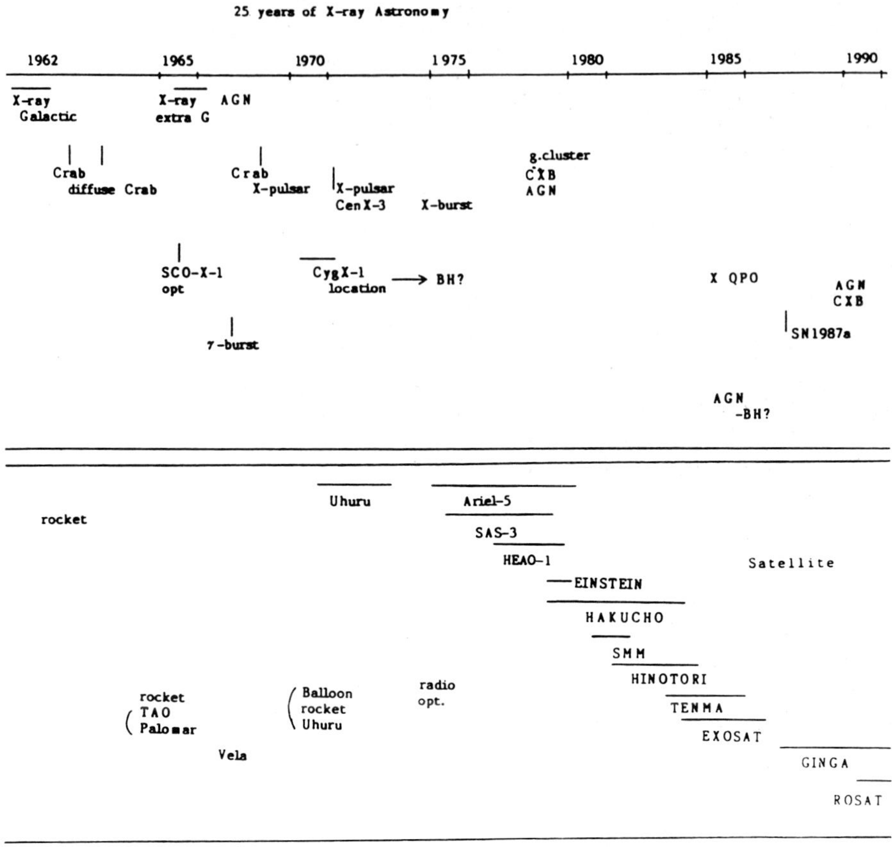

Fig. 9. 25 year history of X-ray astronomy

a

b

c

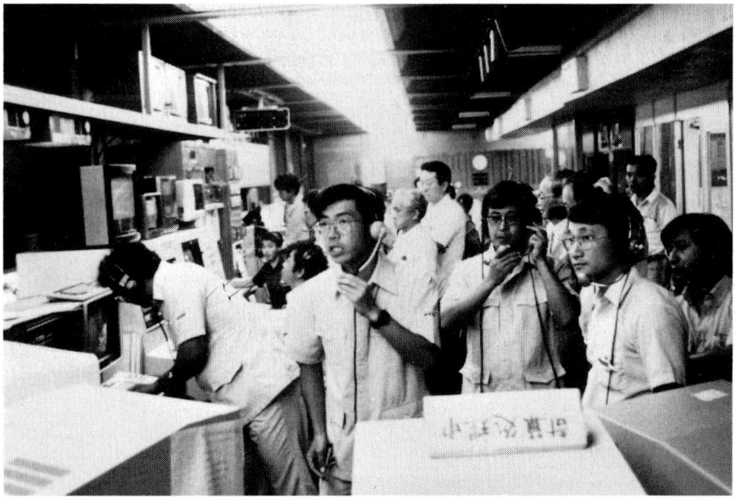

d

Fig.10. a)-d) Launching of the GINGA satellite by the Institute of Space and Astronautical Science.

References

[1] The debate was broadcast via the BBC and its Proceedings were published.

[2] Hayashi,C. (1966) Vol.4 P.171 Ann. Rev. Astron. Astrophys.

[3] Giacconi,R., H.Gursky, F.Paolini and B.Rossi, (1962) Phys. Rev. Lett. 9 439.

[4] Dyson,F. "Infinite in all Directions" (Harper and Row 1988).

[5] Woltjer,L. (1957) B.A.N. 13 301.

[6] Davis, R.Jr., D.S.Harmer and K.C.Hoffman (1968) Phys. Rev. Lett. 20 1205.
Also see Bahcall, J.N. "Neutrino Astrophysics" (Cambridge Univ. Press 1989) and references therein.

[7] Hirata,k.S., T.Kajita and M.Koshiba et al. (1987) Phys. Rev. Lett. 58 1490 and other references also in "Neutrino Astrophysics" [6].
Hirata,K.S. et al. (Kamiokande II collaboration), (1990) Phys. Rev. Lett. 65 1297 and 1301.

[8] Dotani, K., K.Hayashida et al., M.J.L.Turner et al. and A.M.Cruise et al. (1987) Nature 330,230.

[9] Sunyaev,R. et al., C.Reppin et al., E.Kendziorra, A.C.Brinkman et al. and G.K.Skinner et al. (1987) Nature 330,227.

[10] Mather,J.C., E.Cheng et al. (1990) Ap.J.Lett. 354 37.

[11] See e.g. Physics Today "Search and Discovery" (1990) Vol.43 No.10 P.17.

[12] Papers in the Proceedings of 23rd ESLAB Symp. (1989) on X-ray Astronomy, Vol.2. ed.J. Hunt & B.Battrick.

[13] Marshall, F.E., E.A.Boldt et al. (1980) Ap.J. 235 4.
Giacconi, R. and Zamorani, G., (1987) Ap.J. 313, 20.

[14] Joss, P.C. and S.A.Rappaport, (1984) Ann Rev. Astron. Astrophys. 22, 537.

[15] Murakami,T., J.P.Conner, W.D.Evans et al. (1988) Nature 335, 234.

[16] Pelling, R.M. et al. and K.Makishima et al. (1987) Ap.J. 319,416.

[17] Oda, M. (1977) Space Science Reviews, 20 757

[18] Tanaka, Y.Proc. of 23rd ESLAB Symp. (1989) on X-ray Astronomy, Vol.1. ed J.Hunt & Battrick P.3.

[19] Makino, F. and the ASTRO-C Team (1987) Astro. Lett. Commun. 25 223.

Chaos and Complexity: The Results of Non-linear Processes in the Physical World

L.P. Kadanoff

The Research Institutes, The University of Chicago,
5640 So. Ellis Avenue, Chicago, IL 60637, USA

Abstract. The physical world contains an amazing variety of objects with intricate and complex geometrical properties. The laws of physics, on the other hand, are a set of rather simply-stated mathematical rules. So, how does simple law give rise to complex outcome? This paper suggests that complexity results from a natural process of organization in a chaotic environment. Three examples are given to illustrate this point. The first two are mathematical models: one called diffusion limited aggregation (DLA), and the other the avalanche model. The third example is drawn from the world of experimental physics and is convective turbulence.

I. Introduction

In recent decades, there has been a very considerable interest in the question of how complex spatial and temporal structures may arise from simple non-linear dynamical processes. The classic work of Kolmogorov [1] dealt with how a complex pattern of swirls shows up in turbulent hydrodynamic systems while Turing [2] addressed the evolution of structures in biochemical systems. Later on Katchalsky [3] and then Prigogine [4] described the formation of complex structures in dissipative systems. In solids, metallurgists explore a rich variety of different patterns [5-9] all generated dynamically. Dynamically produced objects can also occur in liquids as shown by Saffman and Taylor [10] and later workers [11-14] In this note, I shall describe some recent work on structure formation in chaotic systems with the hope that these considerations might help one guess where the field is going.

In particular, I shall look at three examples of the dynamical formation of complex patterns in space and time.. The first, DLA, or diffusion limited aggregation was invented by Witten and Sander [15] as a mathematical model for the joining of carbon particles in the atmosphere to form pieces of soot.

Springer Proceedings in Physics, Volume 57
Evolutionary Trends in the Physical Sciences Eds.: M. Suzuki and R. Kubo

This discussion will serve as an illustration of complex structure formation and as an introduction to the fractal [16] or scaling [17] concepts which have played a large role in the development of the area. We then turn to dynamical models of avalanches of the type originally proposed by P. Bak, C. Tang and K. Wiesenfeld [18] with the goal of showing how dynamical processes can lead to complex events of many different sizes. The distribution of events is described by the words 'self-organized criticality', which imply that the past events set in motion a chain of processes which will determine and limit the sizes of future events. The final example is the development of structure in convective turbulence as revealed by the experiments of Libchaber and coworkers [19] Here the characteristics of the two previous examples are combined in that there is a complex spatial structure combined with events which are controlled by a feedback mechanism of the 'self-organizing' type.

2. DLA

DLA is a model in which an 'aggregate' sits on a lattice. This aggregate is just a connected group of occupied sites. (See Figure 1). Imagine a particle which comes in from infinity and move from site to neighboring site in the lattice. This particle is termed a 'random walker' since at each step the direction of its motion is picked at random. This step by step progress continues until the walker comes to a site which is next to an already occupied lattice site. At that point, the walker stops and joins the aggregate. In this way, the aggregate grows by one unit. Then another walker is released at infinity and the whole process begins once more. The structure that thereby forms is very light and consists mostly of holes. (See Figure 2)

Despite the randomness of the growth process, and despite the fact that the different structures produced by different realizations of this process may be quite different, nonetheless there are some rather simple and general laws

<u>Figure 2</u> A DLA aggregate. The structure looks like a set of branches which occupy a very small fraction of the available area. Each branch is composed of sub-branches, looking like the main branches. These sub-branches are in turn composed of smaller units which look like the larger ones. This particular picture was produced by Michael Leibig by a process slightly different from the one described in the text.

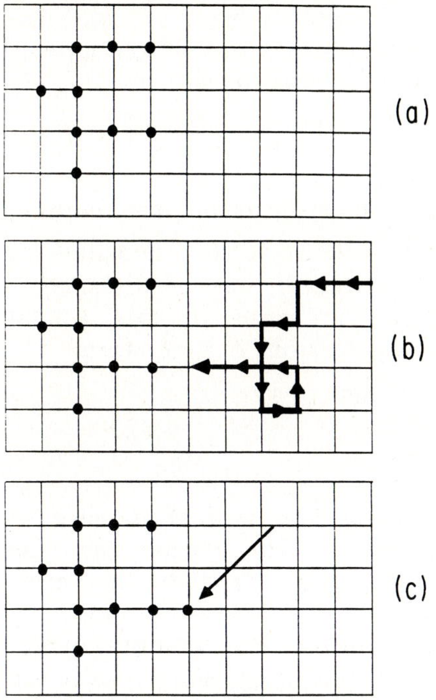

Figure 1 The DLA process. Part a shows the aggregate which occupies a region on the lattice. In Part b, a walker comes in from infinity, chooses a path at random, and comes to a stop when it gets to a site neighboring the aggregate. Part c shows the aggregate increased by one unit. After the walker comes to a stop, then the process starts all over again.

which govern the structures which are outcomes of this growth process. For example, as the mass of the cluster grows, its linear dimension grows as a power of that mass

$$\text{linear dimension} \sim M^x \quad . \tag{1}$$

In two dimensions, x is about 0.6. In this usual terminology, (see Mandelbrot, ref 16) x^{-1} is called the fractal dimension of the aggregate The density of the structure at a distance r from its initial starting point is given in terms of this fractal dimension by a 'scaling function' expression of the form

$$\text{density}(r) = r^{x^{-1}-d} \; \rho\left(\frac{r}{M^x}\right) \tag{2}$$

Notice that the first factor, which gives the typical density in the structure, goes down as r increases. The reason that the structure is so light and filled with holes is that the arms sticking out prevent the random walker from penetrating into the center of the object. Therefore the centers never quite get filled up. Thus the structure adjusts itself in precisely the proper manner so that it will allow itself to be partially filled but not totally filled. To see more of this adjustment process, turn to our next example:

3. Scaling and Universality in Avalanche Models

Bak, Tang, and Wiesenfeld [20] invented a cute dynamical model which shows how richly complicated events can arise in a relatively simple dynamical system. Let model sand [21] be arranged upon a model sandpile, built upon a regular d-dimensional lattice. A one-dimensional version of the model is shown in Figure 3. To start the cascade, one grain of sand is added to one of the columns, picked at random. In between additions, there are cascades of events in which sand falls downhill in response to a too-large local slope of the pile. In the particular model depicted in Figure 3, the sand falls over whenever the column in question stand more than two above its right-hand neighbor. In that case, two grains of sand 'fall over' and land on the two columns to the immediate right of the unstable site. If a grain of sand reaches the right-hand end of the system, it falls off and disappears from view. At any given time, all columns that can fall do so, simultaneously. Then, after this motion if in the new configuration, there are

Figure 3 The Sand Model. This picture shows a short avalanche in progress in a small model sandpile. The first row shows the pile before the addition. The added grain is shown in black in the second row. Unstable grains are shown as boxes. They fall over until a stable configuration is finally reached.

any unstable columns, they fall and so on. Thus the system can sustain 'avalanches'. The cascade of events continues until no more columns are unstable. Then another grain of sand is added at random and the entire process begins once more. The algorithm for the model is shown in the box below:

An Avalanche Model

Square Sand stacked up in a region of size L

A. Add a Grain at a Random Site

(**Avalanche begins**)

B. If the Slope (Height Difference) is greater than 2, two grains from stack fall over.

At right hand end, grains fall off

Continue until no more stacks are unstable. (**Avalanche Ends**)

C. Return to A.

These avalanches in this system can be small or they can cover the entire system many times over. In a recent paper [22],, a group of us studied the nature of the probability distributions $\rho(X|L)$ for the probability that an event of size X will occur in a system with spatial extent L. We looked at two different quantities in some detail:

a. The drop number D. In this case X=D is the number of grains which fall off the end between two addition events.

b. The flip number F. In this case X is the number of falling events which occur between two additions

These model sands slides can be considered to be an example of self-organized criticality. That is, the system organizes itself in such a way that it is marginally stable against the growth of an avalanche. An avalanche event, once started, has a finite probability for growing larger at every stage of its early existence. It also has some probability for growing smaller and dying. These two probabilities balance out so that events of all sizes occur in the sand pile.

It is interesting to ask the reason that the model pile organizes itself in such a fashion. To do this, we must look into the dynamics. Imagine that the pile were relatively flat so that the slides would almost always terminate after a few grains fell. Then, step by step as more sand were added the pile would become steeper and steeper. Conversely, if the pile were very steep, the addition of one grain would trigger a very large event in which many grains would move downhill and make the pile much less steep. The only type of configuration which would not disappear almost at once, is one of an intermediate steepness at which the pile is marginally stable. Thus, in this argument of Bak, Tang, and Wiesenfeld, the marginal stability is a necessary result of the nature of the underlaying dynamics. (Notice incidentally, that the eventual result of a very large avalanche event is that much sand will fall off the end, reducing the slope and thus after-a-while bringing the avalanche to a halt.)

One could use the same kind of argument to describe the configurations produced by the DLA process. These configurations contain very long branches pointing outwards. If these branches were very long and thin, most of the walkers would hit the sides and the branches would fatten up. If, on the other hand, they were very thick, their tip ends would be liable to grow and stick out. As they became more exposed they would stick out further and there would be an increased likelihood that the walkers would land on them. In the end, the fat branches would produce much thinner structures.

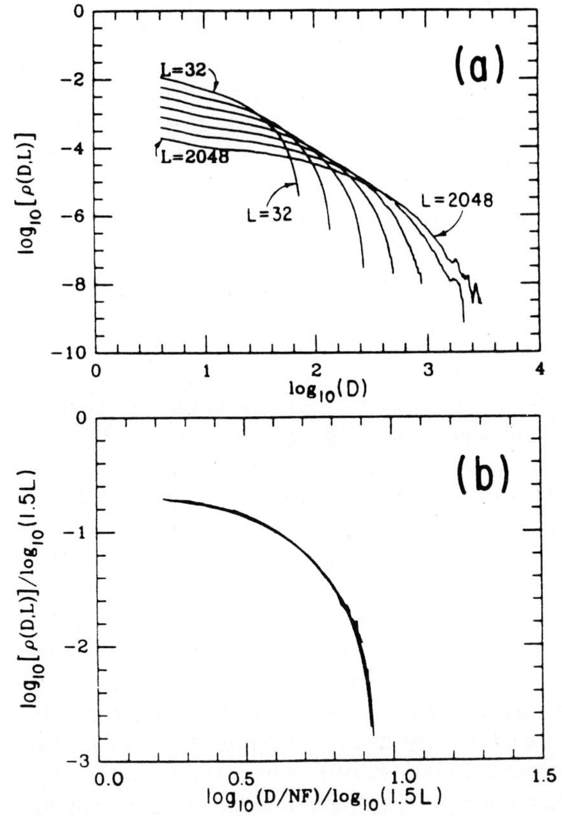

Figure 4. (from Kadanoff, et. al, reference 22. Figure 1.) Distribution of drop numbers in a one-dimensional model. The system size, L, ranges from 32 to 2048. Part a gives the raw data. Part b shows the best fit to these data using a result of the form of equation (3). This fit is very good indeed.

In both cases, the dynamics controls the result. However, in the avalanche model, the net result is not only a geometrical structure (a pile with a certain, marginally stable, slope) but also a distribution of event sizes. The produced distribution is one in which a whole range of events from very small ones to very large one will occur with appreciable probability. Figure 4a shows a probability distribution $\rho(D|L)$. This function describes the probability that in a stack of size L (see Figure 1) there will be an avalanche in which D grains of sand drop off the end. The beginning of the avalanche is defined by the addition of a new grain of sand at random, and the end occurs when all the grains stop moving. Notice that there is a tremendous range of events, from ones in which essentially no grains fall off to ones in which the whole surface layer slides off.

Once again it is interesting to ask whether there is any simple law which fits the observed data. Does this avalanche process give any simple 'laws of nature'. Interestingly enough the answer is apparently 'yes'. All the data for large D and L shown in Figure 4a can be fit by a law of the form

$$\ln \rho(D|L) = f(\alpha)$$
$$\alpha = \frac{\ln D}{\ln L} \quad \text{for } L, D \gg 1 \tag{3}$$

Thus if we plot ln r(D|L)/ln L on the y axis against ln D/ln L on the x axis, for sets of data taken at different values of L, we should expect that all the data would fall on a single curve. Figure 4b shows such a plot. Within the accuracy of the simulational work, all the data does fall onto a curve. Hence Eq (3) does provide a simple 'law' which summarizes the behavior of the avalanches.

It is even more remarkable to notice that this law remains true, with the very same form of the function f(a), even when the model is changed somewhat by, for example, changing the rules by which the sand falls over. This property that the laws remain the same as the model is changed is called 'universality'. It is a characteristic and expected feature of models with self-organizing properties and large-sized events.

4.Turbulence in Thermal Convection

Recent experiments by Albert Libchaber's group [23,24] have shown how very beautiful and intricate structures may arise in situations in which one expects to observe hydrodynamic turbulence. The experiments in question involve a 'Rayleigh-Bénard' flow. In this flow, the fluid is set into motion by placing it in a box and heating it from below. The heated, and therefore less dense, fluid will rise. Correspondingly, the colder and more dense fluid will fall. The net effect that the whole fluid will move and carry heat from bottom to top.

There is a dimesionless number which describes the strength of the forcing in this type of experiment. It is called the Rayleigh number and is given by

$$Ra = \frac{g\alpha\Delta L^3}{\kappa\upsilon} \tag{4}$$

Here, g is the acceleration of gravity, α is the volume thermal expansion coefficient, Δ is the temperature difference between the bottom and the top of the cell, κ and ν are respective the thermal diffusivity and the kinematic viscosity, and L is a characteristic size of the cell. More physically, $g\alpha\Delta$ is a typical size of the buoyant forces which are trying to get fluid into motion while κ and υ are measures of the contrary forces which are working to put the system into equilibrium. High values of Ra correspond to large forcings and indicate a situation in which the fluid is likely to be highly turbulent. The turbulence is easily seen for example (see reference 23) in a setup in which the working fluid is water in a rectangular tank with typical dimension about 20 cm and typical temperature difference Δ about 10 degrees centigrade. In that case, one can see a very chaotic pattern of motion within the tank. Measurements on an analogous system composed of low-temperature helium gas, where there is a better possibility for accurate measurement but a worse possibility for visualization, show scaling types of behavior roughly analogous to the ones discussed above. The environment is certainly very noisy and the measurements show many elements of randomness.

Nonetheless, the observation of the fishtank, as represented in the cartoon view of Figure 5 show considerable structure and regularity. There appear to be at least four different functional regions of the system:

A. Boundary Layer. This thin layer near the top and bottom of the tank contains fluid almost at rest. In these regions, heat is entering the tank via ordinary heat conduction processes.

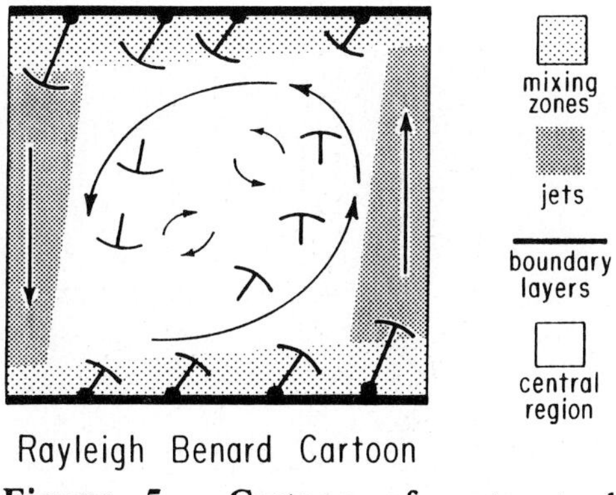

Figure 5. Cartoon of water tank. The bumps on top of the boundary layer are intended to depict the waves. Plumes and thermals are also roughly sketched in.

B. Central Region. This is a relatively quiet region near the center of the tank. The fluid undergoes a gentle regular circular motion together with a more random swirly motion in this region. Hot and cold umbrella-shaped objects called thermals carry heat through this region.

C Mixing Zone. Just above and below the boundary layers, one has a thicker region in which the motion and the temperature difference are much more violent and extreme than in the central region. Here mushroom-shaped structures containing hot (or cold) fluid rise (or fall). These 'plumes' differ from the thermals of the central region in that they are rooted to the boundary layers and have their motion fed from these layers.

D. Jet. Along the side walls there are orderly upwelling on hot fluid on one side and downward motion of cold fluid on the other.

This whole complex machine seems to be a self-regulating structure similar to those discussed in the DLA and avalanche sections. In particular, the amount of plume-like motion seems to regulate itself. There seem to be several parts of this regulatory system. These include: (See Figure 6)

1. The plumes and thermals which move respectively thorough the jets and the central region until they hit the top and bottom, whereupon they produce

2. Boundary layer waves. These are regions of increased thickness in the boundary layer which move rightward in the

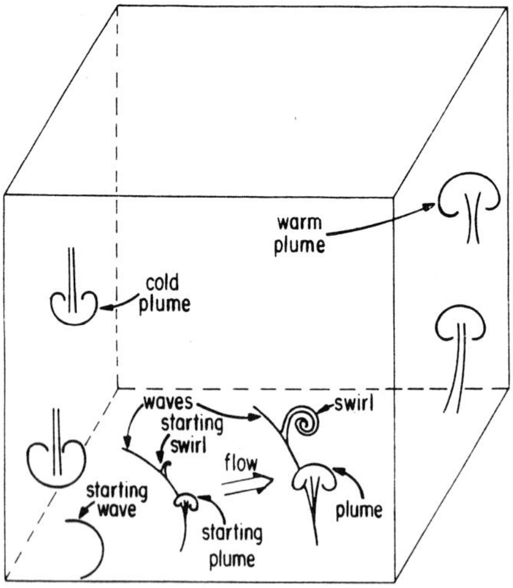

Figure 6. The processes in the water tank. (Taken from Zocchi, Moses, and Libchaber, ref. 23. This is the experimentalists' drawing of the life cycle of plumes.

bottom of the cell of Figure 5 and leftward at the top. Along the top of the bottom boundary layer waves there seem to be

3. Upwellings in the form of spray which then seem to rise and form into plumes.

4. 'Shocks'. In addition, when the boundary layer waves hit the side walls, they apparently produce splashes or shocks, strong temperature disturbances which form an apparently circular pattern and move through the central region. When these hit the boundary layers, they too produce boundary layer waves.

Thus the plumes, thermals, shocks, and waves seem to form a mutually regulating system somewhat like the kind of self-organized criticality discussed earlier in this note.

5. Conclusions

In this note, I have examined several examples in which physicists are attempting to explain observed phenomena by saying that dynamical systems can produce complicated objects via the repetition of simple elementary processes. Once this idea is pointed out it seems simple and inevitable. However, it may be that the organizational idea of Kolmogorov, Turing, Katchalsky , Prigogine, and all their followers is only the beginning of a rich subject of scientific study. We can make assertions about complexity and structure, but we are just beginning to touch the surface of the knowledge which can be gained in this way. One can hope that by following along the path trodden by such leading scientists as Professors Emeritus R. Kubo and H. Mori and, for example, Professors Y. Kuramoto, K. Kawasaki, M. Suzuki, K. Kaneko, Y. Sawada and many others, we shall begin to gain some understanding of how the beautiful simplicity of physical law can lead to the beautifully intricate structure of the world about us.

Acknowledgement

Some of the research described here was supported by the University of Chicago MRL. Many of the ideas reported here have been discussed in the interactive and collaborative atmosphere provided by that Laboratory and by the Research Institutes of the University of Chicago.

References

[1] A. N. Kolmogorov, CR. Acad. Sci. USSR **32**, 16 (1941). A. N. Kolmogorov, J. Fluid Mech. 13, 82 (1962).

[2] A.M. Turing, Phil. Trans. Roy. Soc. London B 273, 37 (1957).

[3] A. Katchalsky and P.F. Curan, Nonequilibrium Processes in Biophysics (Harvard University Press, Cambridge, Mass., 1967)

[4] G. Nicolis and I Prigogine Self-Organization in Nonequilibrium Systems (John Wiley, New York, 1977)

[5] R.T. Delves. "Theory of Interface Stability" in Crystal Growth ed. B. Pamplin (Pergamon, Oxford, 1975). B Chalmers, Principles of Solidification (John Wiley, New York, 1964).

[6] R. Doherty, Dendritic Growth, in Crystal Growth ed. B. Pamplin (Pergamon, Oxford, 1975).

[7] W. W. Mullins and R.F. Sekerka, J. Appl, Phys. **34**, 323 (1963).

[8] J.S. Langer, Rev. Mod. Phys. **52**, 1 (1981), Lectures in the Theory of Pattern Formation, from Les Houches Summer School 1986.

[9] D. Kessler, J. Koplik and H. Levine, Phys. Rev A **32**, 1930 (1985), R. Brady and R.C. Ball, Nature (London) 309, 225 (1984).

[10] G.I. Taylor and P.G. Saffman, Q.J. Mech. Appl. Math. **12**, 265(1959)

[11] D. Grier, E. Ben-Jacob, Roy Clark, and L.M. Sander, Phys. Rev. Lett. **56**, 1264 (1986). Y. Sawada, A. Daugherty, J.P. Gollub. Phys. Rev. Lett. **56**, 1260 (1986).

[12] Shoudan Liang, Phys. Rev. A **33**, 2663 (1986). J. Kertesz, et. al. in On Growth and Form (Ed. H.E. Stanley and N. Ostrowshy, Martinus Nijhoff, Boston, 1986) p. 249.

[13] D. Bensimon, L.P. Kadanoff, S. Liang, B. Shraiman, C. Tang, Rev. Mod. Phys. October 1986. D. Bensimon, L.P. Kadanoff, S. Liang, B. Shraiman, C. Tang, Directions in Condensed Matter Physics, (Ed. G. Grinstein, G. Mazenko, World Scientific, Singapore, 1986) p. 51. D. Bensimon, L.P. Kadanoff,

S. Liang, B. Shraiman, C. Tang, Revista Mexicana de Fisica (suplemento) 32, S101 (1986).

[14] G.T. Dee, Physica **15D**, 295 (1985). J. Stat. Phys. **39**, 705 (1985).W. van Saarlos, Phys. Rev. A 37, 211 (1988).J.S. Langer and L.A. Turski, Acta Metall. **25**, 1113 (1977).J.S. Langer,Rev. Mod. Phys.} 52, 1 (1980)..
R.C. Brower, D.A. Kessler, J. Koplik, and H. Levine, Phys. Rev. **A 29**, (1984),E. Ben-Jacob, N. Goldenfeld, J.S. Langer, and G.Schön, Phys. Rev. **A 29,** (1984).

[15] T.A. Witten and L.M. Sander, Phys. Rev. B 27, 5696 (1983)

[16] B Mandelbrot The Fractal Geometry of Nature (W. H. Freeman 1982).

[17] See, for example, Leo Kadanoff Physica A **163** 1 (1990). for a review.

[18] P. Bak, C. Tang and K. Wiesenfeld, Phys. Rev. Lett. **59**, 381 (1987); P. Bak, C. Tang and K. Wiesenfeld, Phys. Rev. **A 38**, 364 (1988)

[19] B. Castaing, G. Gunaratne, F. Heslot, A. Libchaber, Leo Kadanoff, S. Thomae, Xiao-zhong Wu, S. Zaleski and G. Zanetti JFM **204**,1 (1989). Xiao-Zhong Wu, Leo Kadanoff, Albert Libchaber, and Masaki Sano, Phys. Rev. Letts **64** 2140 (1990).

[20] P. Bak, C. Tang and K. Wiesenfeld, Phys. Rev. Lett. 59, 381 (1987); P. Bak, C. Tang and K. Wiesenfeld, Phys. Rev. A 38, 364 (1988)

[21] Real sand does not behave in precisely the same manner as model sand. See H.M.Jaeger, Chu-heng Liu and Sidney R. Nagel, Phys. Rev. Lett. **62**, 40 (1989) but also G.A.Held, D.H. Solina II, D.T. Keane, W.J..Haag, P.M. Horn and G. Grinstein IBM Yorktown Heights preprint (1990).

[22] Leo Kadanoff, Sidney Nagel, Lei Wu and Su-min Zhou, Phys Rev. **A39** 6524 (1989)

[23] Giovanni Zocchi, Elisha Moses, and Albert Libchaber Physica A 166 387 (1990).

[24] S. Gross, G. Zocchi and A. Libchaber Compt. Rend, Acad Sci (Paris) 307 (1988). E. Moses, G. Zocchi, I. Procaccia, and A. Libchaber (preprint 1990).

New Trends in the Physics of Phase Transitions

M. Suzuki

Department of Physics, Faculty of Science,
University of Tokyo, Bunkyo-ku, Tokyo 113, Japan

Abstract

A brief historical review of theories of phase transitions and critical phenomena is presented with emphasis on basic concepts and with even more emphasis on new trends. In particular, a general approach to critical phenomena, the so-called coherent-anomaly method (CAM) is explained together with some applications to exotic phase transitions such as spin glasses, chiral orders and the KT-transition. The CAM is based on the generalized effective-field theory which is constructed in terms of Kubo's canonical correlations. A possible application of the CAM to high-T_c superconductivity is discussed.

1 Introduction

The study of phase transitions and critical phenomena has made great progress in the four decades since Dr.Y.Nishina died. In 1873, before Dr.Nishina was born, van der Waals discovered the equation of state for the gas-liquid transition and in 1907 P.Weiss [1] proposed the first mean-field theory to study phase transitions self-consistently. His theory is very simple but his idea of mean fields is still useful in many-body problems. In 1925, E.Ising [2] solved the linear Ising chain, which is the first exact solution of spin statistics. In 1944, Onsager [3] solved the two-dimensional Ising model to show the logarithmic singularity of the specific heat near the critical point. This was an epoch-making work in the field of phase transitions. Ten years after Onsager's success, Yang [4] derived the exact expression for the spontaneous magnetization of the two-dimensional Ising model on the square lattice, which had been already announced by Onsager without any derivation. In 1971, Baxter [5] solved the eight-vertex model, whose solution is quite remarkable in the sense that the critical exponents of it vary continuously as functions of interaction strength. This violates the universality [6,7] of critical phenomena, but satisfies the weak universality [8]. In 1970, Polyakov [9] proposed conformal field theory. This is now found to be useful in understanding many different universality classes of critical phenomena in two dimensions. Unfortunately it is inapplicable to higher dimensions. Recently Howes, Kadanoff and den Nijs

Springer Proceedings in Physics, Volume 57
Evolutionary Trends in the Physical Sciences Eds.: M. Suzuki and R. Kubo

[10], and H.Au-Yang et al. [11] studied the chiral Potts model, which has a new feature of genus higher than unity.

On the other hand, Landau [12] proposed his famous phenomenological theory of second-order phase transitions. This is still a standard theory of phase transitions. A modern treatment of critical phenomena started when Fisher [13] proposed his asymptotic scaling form of a correlation function $C(R)$ as

$$C(R) \sim \frac{1}{R^{d-2+\eta}} \exp(-\frac{R}{\xi}) \tag{1.1}$$

where R denotes the distance of the two relevant stochastic or quantum variables such as spins, and ξ the correlation length. Here, d denotes the dimensionality of the system and η is so-called Fisher's exponent which expresses the deviation of $C(R)$ from the Ornstein-Zernike form. In 1965, Widom [14] proposed the following homogeneity law of the equation of state, namely

$$m = \varepsilon^{\beta} f(h/\varepsilon^{\Delta}) \tag{1.2}$$

where m is an order parameter, say, magnetization in ferromagnets, and h denotes an external field conjugate to the order parameter m. Here $\varepsilon = (T - T_c^*)/T_c^*$, where T_c^* denotes the true critical point. This leads to the scaling relations of critical exponents such as $\alpha + 2\beta + \gamma = 2$, where $\alpha; \beta$, and γ denote the critical exponents of the specific heat, spontaneous magnetization and susceptibility, respectively. For example, the susceptibility χ_0 has the singularity of the form $\chi_0 \sim \varepsilon^{-\gamma}$ near the true critical point T_c^* . This homogeneity ansatz was also proposed independently in England, USSR and Japan. Such a situation happens some times in science. Kadanoff [15] gave a clear physical picture to this homogeneity ansatz, namely the cell analysis approach to the scaling law. In 1971, Wilson [16] formulated the renormalization group approach to critical phenomena. This is a microscopic theory of the scaling law to give explicitly the values of critical exponents. In 1972, Fisher [17] proposed the finite-size scaling theory, by which the coherent-anomaly method [18] was inspired.

Up to 1972, many simple uniform systems had mainly been investigated. In 1973, Kosterlitz and Thouless [19] found a new type of phase transition without long-range order. This is associated with the condensation of vortex pairs in the plane-rotator model of a ferromagnet. This transition is characterized by the divergence of the susceptibility. Thus, this type of phase transition may be called "a response-diverging transition". In 1975, Edwards and Anderson [20] proposed a theory of spin glasses, in which the concept of frustration [21] plays an essential role and consequently there exist many degenerated ground states. In 1977, Villain [22] emphasized the importance of chirality or chiral order [23,24] in spin systems with competing interactions.

Thus, the year 1973 is a turning period from simple uniform orders to complex topological orders. The super-effective field theory [SEFT] was formulated by the present author [25] in order to study such exotic phase transitions as spin glasses, chiral orders and other complex topological orders. This SEFT is particularly useful in strongly interacting systems. The combination of the CAM and SEFT gives a unified theory of phase transition, as will be seen in the present paper.

As a summary of exactly solvable systems, we propose here a tree diagram of solvability relationship in phase transitions, as shown in Fig.1. Here, the symbol " ⓐ → ⓑ " denotes that if ⓐ is solved, then ⓑ is solved. The notation " ⓐ ↔ ⓑ " means that both are equivalent to each other. For example, the one-dimensional XY-model was found [26] to be equivalent to the two-dimensional Ising model. In more general terms, it was proven by the present author [27] that a d-dimensional quantum system is transformed into the corresponding $(d+1)$ - dimensional classical system. This equivalence theorem is the general basis of the quantum Monte Carlo method [27, 28].

As an example to explain the frontier of investigation shown in Fig.1, we propose here an extended IRF model, in which each elementary cell is given by an $m \times n$ cell with all kinds of interaction inside each cell. Then, there may exist many exactly solvable cases which can be hopefully found by solving the corresponding extended Yang-Baxter relation.

In connection with these exactly solvable systems, it should be remarked here that the two-dimensional Ising model with a specific imaginary magnetic field $h = \beta\mu_B H = \frac{1}{2}i\pi$ solved by Lee and Yang [29] is found [30] to be equivalent to Villain's fully frustrated model. This suggests a possibility [30] to study frustrated systems using the corresponding Lee-Yang systems with an imaginary magnetic field.

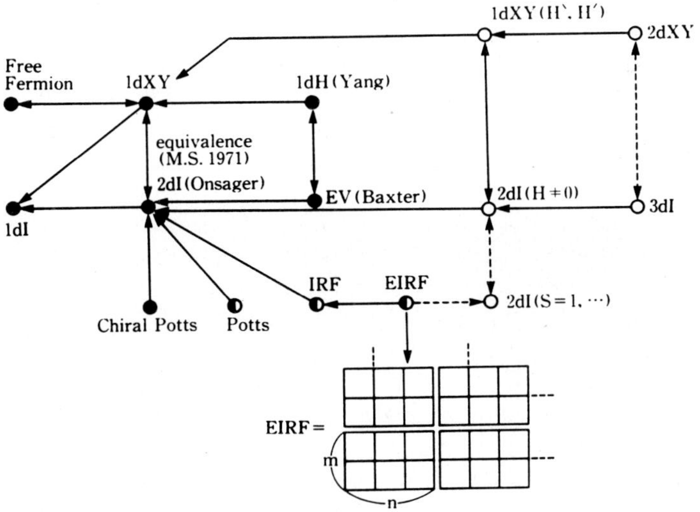

Fig.1. Tree diagram - solvability relationship. Here the symbol " ● "denotes "solved", " ○ " , unsolved, and " ◐ ", partially solved.

2 Basic Concepts of Phase Transitions

It will be instructive to summarize here basic concepts of phase transitions. It should be emphasized first that they occur only in the thermodynamic limit, namely in the infinite limit of the system size L (i.e., $L \to \infty$).

They are characterized by the appearance of Yang's ODLRO (namely off-diagonal long-range order) [31]. It is defined by the following decoupling properties of the correlation functions

$$\langle \psi^\dagger(0)\psi(R)\rangle \to \Psi_1^*(0)\Psi_1(R)$$
$$\langle \psi_1^\dagger(0)\psi_2^\dagger(\delta)\psi_3(R)\psi_4(R+\delta)\rangle \to \Psi_2^*(0)\Psi_2(R) \tag{2.1}$$

for a finite δ and for large R.

From a practical point of view, such a picture of emerging order for increasing system size L as shown in Fig.2 is extremely important. This mechanism of emerging order is the basis of the occurrence of broken symmetry. Namely, the boundary effect is enhanced through fluctuating correlations inside the cluster at low temperatures. Thus, the correlations or Kubo's canonical correlations [32] between the center and boundary of the cluster play an essential role in studying phase transitions theoretically. In fact, the mean-field and effective-field theories are all based on this mechanism of emerging order, as will be seen in the succeeding sections.

The homogeneity law of the equation of state (1.2) is derived from the following scaling invariance for the magnetization $m = f(\varepsilon, h)$:

$$f(b^y\varepsilon, b^x h) = b^{d-x} f(\varepsilon, h) \tag{2.2}$$

for the scale factor b with some appropriate scaling exponents x and y. The solution of (2.2) is given in the form (1.2) with $\beta = (d-x)/y$ and $\Delta = x/y$.

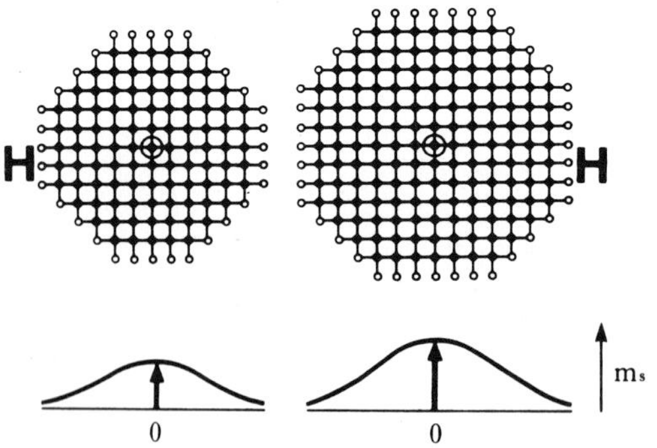

Fig.2 The order parameter at the center O emerges below the critical point, when the system size L increases. Conversely, the order parameter decreases above the critical point as L increases.

Fisher's finite-size scaling law [17] has also been used successfully in estimating critical exponents numerically. This is formulated as follows. The total magnetization in a system with size L takes the form

$$m(L) \simeq L^{d-\beta/\nu} F(\frac{h}{\varepsilon^{\Delta}}, \frac{L}{\xi}) \tag{2.3}$$

for the correlation length ξ of the form $\xi \sim \varepsilon^{-\nu}$.

3 Generalized Mean-Field Theory

In our new approach [18] to critical phenomena, the mean-field approximation and its generalization play a substantial role, as will be seen in the succeeding sections. Thus, we start with a brief review of the Weiss mean-field theory [1] and the Bethe approximation [33].

As is well known, the susceptibility of a ferromagnet obtained in the Weiss mean-field approximation is given by

$$\chi_0^{(mf)} \simeq \frac{1}{\varepsilon}\bar{\chi}^{(mf)}; \varepsilon = (T - T_c^{(mf)})/T_c^{(mf)}, \tag{3.1}$$

and we have also a similar result with a different coefficient $\bar{\chi}^{(B)}$ in the Bethe approximation. Here, it should be remarked that $\bar{\chi}^{(B)} \gg \bar{\chi}^{(mf)}$. This observation of a drastic change of the coefficients for increasing degree of approximation suggests our new approach, namely the coherent-anomaly method [18,34-41].

Now we consider a generalized cluster Ω shown in Fig.3. The corresponding effective Hamiltonian of this cluster Ω is given by

$$\tilde{\mathcal{H}} = \mathcal{H}_{\Omega} - M_{\partial\Omega} H_{\text{eff}} - M_{\Omega} H, \tag{3.2}$$

where $\mathcal{H}_{\Omega}$ denotes the original Hamiltonian of the cluster Ω and $\partial\Omega$ denotes

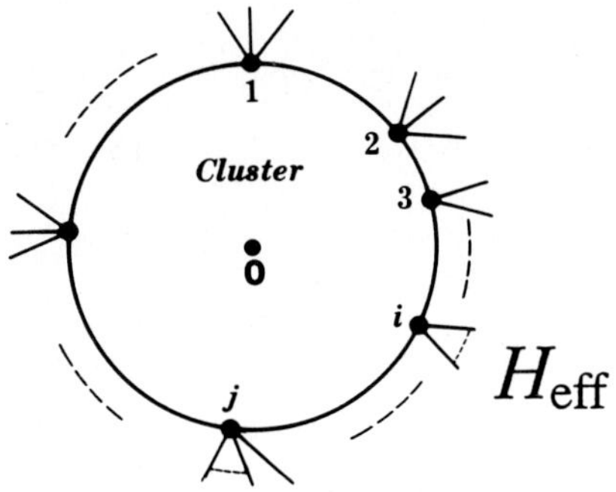

Fig.3 A generalized cluster with mean fields on the boundary [18].

the boundary of the cluster. The operator M_Ω is denoted by the sum of spin operators S_j inside the cluster Ω. The order parameter $m_0 = \langle S_0 \rangle$ at the center of the cluster is expressed as

$$m_0 = \langle S_0; M_{\partial\Omega} \rangle \beta H_{\text{eff}} + \langle S_0; M_\Omega \rangle \beta H \tag{3.3}$$

as far as the linear terms with respect to H_{eff} and H are concerned. Here, $\langle A; B \rangle$ denotes Kubo's canonical correlation defined by

$$\langle A; B \rangle = \frac{1}{\beta} \int_0^\beta \langle AB(i\hbar\lambda) \rangle d\lambda \tag{3.4}$$

with $\beta = 1/k_B T$ and

$$B(z) = \exp(\frac{iz}{\hbar} \mathcal{H}_\Omega) B \exp(-\frac{iz}{\hbar} \mathcal{H}_\Omega). \tag{3.5}$$

We assume that the effective field H_{eff} is given by $H_{\text{eff}} = (\hat{z}J)\langle S_j \rangle$, where $\hat{z}$ denotes the number of nearest neighbours outside the cluster at the boundary. Then, we have [18,34,35]

$$m_0 = \mathcal{K}_L(T) m_j + \chi_\Omega(T) H \tag{3.6}$$

where

$$\mathcal{K}_L(T) = \hat{z}\beta J \langle S_0; M_{\partial\Omega} \rangle \qquad \text{and} \qquad \chi_\Omega(T) = \beta \langle S_0; M_\Omega \rangle. \tag{3.7}$$

The quantity $\mathcal{K}_L(T)$ denotes the feed - back effect of the boundary and χ_Ω the finite-size response function. The self-consistency condition is given by $m_0 = m_j (= m)$. Thus we obtain

$$m = \frac{\chi_\Omega(T)}{1 - \mathcal{K}_L(T)} H \equiv \chi_0(T) H. \tag{3.8}$$

The critical point T_c is determined from the pole of the susceptibility $\chi_0(T)$ defined by (3.8), namely $\mathcal{K}_L(T_c) = 1$. Thus, we arrive at the Curie-Weiss law of the form [18,34,35]

$$\chi_0(T) \simeq \frac{T_c \bar{\chi}(T_c)}{T - T_c} \quad ; \quad \bar{\chi}(T_c) = -\frac{\chi_\Omega(T_c)}{T_c \mathcal{K}'_L(T_c)} \tag{3.9}$$

near the approximate critical point T_c. It should be remarked here that the mean-field critical coefficient depends on the approximate critical point T_c.

4 Coherent-Anomaly Method

4.1 Basic Scheme

As was derived in the preceding section, the susceptibility $\chi_0(T)$ shows the classical singularity (3.9) in any generalized cluster- mean-field approximation. In this sense, any extended mean-field approximation has been believed to be useless in studying modern criticality. However, we now find [18] that it is not the situation, and that the above systematic mean-field approximations can afford information on the non-classical critical behaviour of the relevant system. It is shown [18,34] using Fisher's asymptotic scaling correlation function (1.1), that the mean-field critical coefficient $\bar{\chi}(T_c)$ diverges systematically as the degree of approximation increases, i.e., as T_c approaches the true critical point T_c^*, namely

$$\bar{\chi}(T_c) \to \infty \qquad \text{as} \qquad T_c \to T_c^*. \tag{4.1}$$

This systematic divergence is called "the coherent anomaly" [18]. Then, we assume that

$$\bar{\chi}(T_c) \simeq \frac{A}{(T_c - T_c^*)^{\psi}} \tag{4.2}$$

near $T_c = T_c^*$. This form can also be derived from the formulation (3.9) with (3.7) using (1.1), as far as the dominant singularity is concerned. The critical exponent γ defined in

$$\chi_0(T) \sim \frac{1}{(T - T_c^*)^{\gamma}} \tag{4.3}$$

is shown [18,34] to be related to the coherent-anomaly exponent ψ as

$$\gamma = 1 + \psi. \tag{4.4}$$

This is the basic scheme of the coherent-anomaly method to estimate critical exponents using generalized systematic mean-field approximations. This scheme is easily extended to any other physical quantities such as the spontaneous magnetization and correlation length [37].

More intuitively, this scheme can be understood using the envelope of the susceptibilities obtained by systematic mean-field approximations, as shown in Fig.4. This envelope expresses the common feature characteristic of each mean-field result.

4.2 Transfer-Matrix CAM Theory

A typical systematic series of mean-field approximations is constructed using the transfer-matrix method. That is, we consider strip systems with infinite length in one direction and finite width in other directions. We apply a mean-field or effective fields at the boundary of the strip systems. These systems can be solved using the transfer-matrix method.

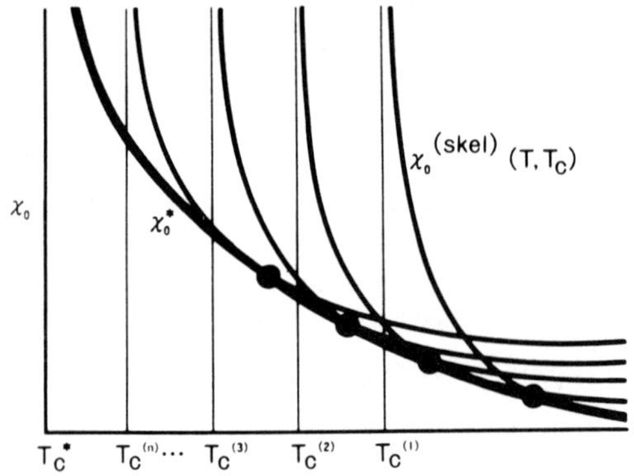

Fig.4 Envelope of the mean-field approximations [18]. Here, χ_0^* denotes the envelope corresponding to the true singularity and "skel" means skeletonization.

For example, the two-dimensional Ising model has been studied by Hu et al. [36] using this transfer-matrix CAM theory to give the result $\gamma = 1.749$ only for three systematic strips, namely 3, 5 and 7- line strips . This shows how useful the CAM theory is in estimating critical exponents accurately.

4.3 Multi-Effective-Field CAM Theory

Quite recently the present author [40] proposed a general theory of multi-effective-field approximations. We now consider such a cluster shown in Fig.2. The effective Hamiltonian of the cluster is composed of the following two parts :

$$\mathcal{H}_{\text{eff}} = \mathcal{H}_{\Omega} + \mathcal{H}_{\partial\Omega} \tag{4.5}$$

where $\mathcal{H}_{\Omega}$ denotes the original interaction of the cluster and $\mathcal{H}_{\partial\Omega}$ may be written in the form

$$-\beta \mathcal{H}_{\partial\Omega} = \sum_{j\epsilon\partial\Omega} K_j^{(1)} S_j + \sum_{i,j\epsilon\partial\Omega} K_{ij}^{(2)} S_i S_j + \cdots + \sum_{j_1,j_2,\cdots j_n\epsilon\partial\Omega} K_{j_1,j_2,\cdots j_n}^{(n)} S_{j_1} S_{j_2} \cdots S_{j_n} + \cdots \tag{4.6}$$

for example, in Ising spin systems. Some restricted number of effective fields can be determined from the self-consistency conditions

$$\langle S_{i_1} S_{i_2} \cdots S_{i_n} \rangle = \langle S_{j_1} S_{j_2} \cdots S_{j_n} \rangle \tag{4.7}$$

Since there are many combinations of multi-effective-fields for a fixed cluster, it is possible to construct a certain systematic series of effective-field approximations even for a fixed cluster.

For example, Minami, Nonomura, Katori, and the present author [41] studied the 3×3 and 4×4 clusters of the Ising model in two dimensions. From the CAM plot for the 3 × 3 cluster with some appropriate combinations of multi-effective-fields, we have obtained $\gamma = 1.7498$. This is almost exact ($\gamma = 7/4 = 1.75$)! For more details, see Ref. 41.

4.4 Double-Cluster Approximation for the CAM

Quite recently Lipowski [42] emphasized the usefulness of the double-cluster approximation [43-46] in the CAM. In this approximation, we consider two clusters with a size Ω and another size Ω' very close to Ω. Let s be an order parameter of the relevant spin system :

$$s = \frac{1}{\Omega} \sum_{j \in \Omega} S_j . \tag{4.8}$$

The self-consistency condition is given by $\langle s \rangle_\Omega = \langle s \rangle_{\Omega'}$. The average $\langle s \rangle_\Omega$ can be expanded as

$$\langle s \rangle_\Omega = f_\Omega(T) H_{\text{eff}} + \chi_\Omega(T) H \tag{4.9}$$

as far as the linear terms in H_{eff} and H are concerned. The average $\langle s \rangle_{\Omega'}$ is also expanded similarly. Thus, we obtain

$$\chi_0(T) \sim \widehat{H}_{\text{eff}} \; ; \; \widehat{H}_{\text{eff}} = \frac{\chi_{\Omega'}(T) - \chi_\Omega(T)}{f_\Omega(T) - f_{\Omega'}(T)} \tag{4.10}$$

with $H_{\text{eff}} = \widehat{H}_{\text{eff}} \cdot H$. The critical point T_c is determined by $f_\Omega(T_c) = f_{\Omega'}(T_c)$. This equation is shown phenomenologically to have a unique critical point as follows. At low temperatures ($T \ll T_c^*$), we have $\langle s \rangle_\Omega < \langle s \rangle_{\Omega'}$ for $\Omega < \Omega'$ and at high temperatures ($T \gg T_c^*$) we have $\langle s \rangle_\Omega > \langle s \rangle_{\Omega'}$ for $\Omega < \Omega'$. Thus, there exists a critical point in this double-cluster approximation. This double-cluster CAM is now being applied successfully to many problems.

4.5 CAM Theory at the Ground State

It is possible to extend the CAM to systems at $T = 0$. The eigenvalue problem $\Psi_g = E_g \Psi_g$ for the ground state Ψ_g is replaced by the effective eigenvalue problem

$$\mathcal{H}_{\text{eff}} \Psi_\Lambda = E_\Lambda \Psi_\Lambda \; ; \; \mathcal{H}_{\text{eff}} = \mathcal{H}_\Omega - \sum_{j \in \partial\Omega} \Lambda_j Q_j \tag{4.11}$$

for any possible order parameter Q_j. The self-consistency condition is again expressed by the canonical correlation

$$\lim_{\beta \to \infty} \beta \langle A; B \rangle_g = \lim_{\beta \to \infty} \int_0^\beta \langle A B(i\hbar\lambda) \rangle d\lambda \tag{4.12}$$

which requires calculations of excited states of the cluster Hamiltonian $_\Omega$.

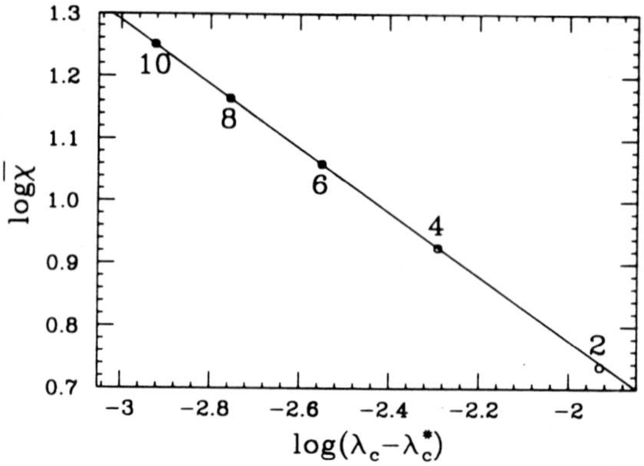

Fig.5 CAM plot of the susceptibility in the $1D$ XY model.

Quite recently Nonomura and the present author [47] studied the critical beh avior of the one-dimensional XY model defined by the Hamiltonian

$$\mathcal{H} = -\sum (\sigma_j^x \sigma_{j+1}^x + \lambda \sigma_j^y \sigma_{j+1}^y) - H \sum \sigma_j^y \tag{4.13}$$

at the ground state ($T = 0$) . This system shows a phase transition at $\lambda = \lambda_c^* = 1$. Using the ground state CAM theory, we[47] have obtained $\gamma = 1.4996$, which is extremely close to the expected exact value $\gamma = \nu(2 - \eta) = 1 \cdot (2 - 1/2) = 3/2 = 1.5$ obtained by McCoy [48]. An example of the CAM plot for this system is shown in Fig.5.

4.6 Dynamical CAM Theory

The CAM theory is also extended to dynamics [18,38]. We consider the following dynamical cluster mean-field approximation, in which our dynamical effective Hamiltonian takes the form

$$\widetilde{\mathcal{H}}(t) = \mathcal{H}_\Omega - M_{\partial\Omega} \cdot H_{\text{eff}}(t) - M_\Omega \cdot He^{i\omega t}. \tag{4.14}$$

The density matrix of this effective cluster Hamiltonian obeys the equation

$$i\hbar \frac{\partial}{\partial t} \rho(t) = [\widetilde{\mathcal{H}}(t), \rho(t)]. \tag{4.15}$$

Following Kubo [32], we put

$$\rho(t) = \rho_{eq} + \Delta\rho(t). \tag{4.16}$$

In our case, $\Delta\rho(t)$ is composed of the two parts of the external field and effective-field terms. Namely we have

$$i\hbar \frac{\partial}{\partial t} \Delta\rho(t) = [\mathcal{H}_\Omega, \Delta\rho(t)] - [M_{\partial\Omega}, \rho_{eq}] H_{\text{eff}}(t) - [M_\Omega, \rho_{eq}] He^{i\omega t}. \tag{4.17}$$

as far as the linear terms in $H_{\text{eff}}(t)$ and H are concerned. Then the solution of (4.17) is given by

$$\Delta\rho(t) = \frac{1}{i\hbar}\int_{-\infty}^{t} U(t-t')[M_{\partial\Omega}, \rho_{eq}]H_{\text{eff}}(t') + [M_{\Omega}, \rho_{eq}]He^{i\omega t}U^{-1}(t-t')dt' \tag{4.18}$$

with

$$U(t-t') = \exp[i(t-t')\mathcal{H}_{\Omega}/\hbar]. \tag{4.19}$$

The magnetization $m(t)$ above the critical point is expressed by

$$m(t) = \mathrm{Tr}S_0\Delta\rho(t). \tag{4.20}$$

If we impose the self-consistency condition $H_{\text{eff}}(t) = \hat{z}Jm(t)$, then we arrive at the result

$$\chi(\omega) = \frac{\chi_{\Omega}(\omega)}{1-\mathcal{F}(\omega)}\ ; \ \mathcal{F}(\omega) = \hat{z}J\chi_{\partial\Omega}(\omega), \tag{4.21}$$

and

$$\chi_{\Omega}(\omega) = \lim_{\varepsilon\to+0}\int_0^{\infty}\phi_{\Omega}(t)e^{-i\omega t-\varepsilon t}dt \tag{4.22}$$

with Kubo's response function $\phi_{\Omega}(t)$ defined

$$\phi_{\Omega}(t) = -\frac{d}{dt}\int_0^{\beta}\langle M_{\Omega}(i\hbar\lambda)M_0(t)\rangle d\lambda. \tag{4.23}$$

Similarly $\chi_{\partial\Omega}(\omega)$ is defined in terms of $M_{\partial\Omega}$ instead of M_{Ω} in (4.23).

Near the critical point, the dynamical susceptibility takes the following form

$$\chi(\omega) = \frac{\bar{\chi}(T_c)}{\varepsilon + i\omega\bar{\tau}(T_c)} \quad ; \quad \varepsilon = \frac{T-T_c}{T_c}, \tag{4.24}$$

for small ε and ω in the present approximation. Thus, the relaxation time of the magnetization is given by $\tau = \bar{\tau}(T_c)/\varepsilon$ in our dynamical mean-field approximation. The critical exponent Δ defined by $\tau \sim \varepsilon^{-\Delta}$ is now estimated through the coherent anomaly

$$\bar{\tau}(T_c) \sim (T_c - T_c^*)^{-(\Delta-1)}. \tag{4.25}$$

As an application of the above general formulation, Katori and the present author [38] studied the two-dimensional stochastic model [49,50] to obtain the result $\Delta = 2.15 \pm 0.02$. This is very close to the value $\Delta = 2.13$ obtained by the Monte Carlo method [51]. It should be remarked here that the deviation of the critical slowing down exponent Δ from the susceptibility exponent γ, namely the inequality $\Delta > \gamma$ was first discovered by Yahata and the present author [52] using the dynamical high-temperature expansion method.

5 Generalization of the CAM and Unification of Theories of Critical Phenomena

In the preceding section, the basic scheme of the CAM has been presented. It is easily extended in many directions.

5.1 Power Series CAM

The traditional high-temperature expansions can be analyzed using the concept of coherent anomaly. Namely, they can be regarded as some partial sums of certain mean-field approximations [53]. For example, the inverse of the relevant physical quantity is approximated as a series of polynomials and the zeros of these polynomials play the role of mean-field critical points. The residues of the relevant physical quantity at these poles have been found [53] to show a coherent anomaly, as the poles approach the true critical point.

5.2 Continued-Fraction CAM

It is also possible to construct continued-fraction expansions from given power-series expansions of the relevant physical quantity and to apply [54] the CAM theory to such continued-fraction expansions. These power-series CAM and continued-fraction CAM can be extended to Pade approximations.

5.3 Generalized CAM

In some systematic series of approximations, we define a degree of approximation by the difference between a certain characteristic temperature T_c and the true critical point T_c^*, namely by $(T_c - T_c^*)$. Here, it should be emphasized that T_c is now not necessarily a critical point, but an arbitrary characteristic temperature which approaches systematically (or coherently) the true critical point.

For example, we study the singularity of the susceptibility of a ferromagnet using a systematic series of finite lattices with periodic boundary condition. We may take as T_c the temperature at which the specific heat of the corresponding finite lattice or the slope of the susceptibility becomes maximum. Then the susceptibility $\chi(T_c)$ shows a coherent anomaly as T_c approaches T_c^*.

5.4 Unification of Various Approximate Theories of Critical Phenomena

As is seen from the above arguments, the CAM theory gives a unified picture or treatment of critical phenomena through the concept of coherent anomaly. In this sense, the CAM theory is a "metapproximation" or analytic continuation of approximate theories.

We may also say that analytic theories of phase transitions are gradually merged into large-scale numerical calculations in the CAM theory. For example, the Weiss mean-field theory is the simplest but most useful approximation of phase transitions. Over these last eighty years, there have been many papers published on phase transitions only using the Weiss approximation, but in our CAM theory we have to construct a systematic series of generalized mean-field approximations which requires larger-scale calculations for clusters in order to take larger fluctuations into account. In this sense, the CAM theory is a unification of analytic theories and computational physics, and consequently it is found to be appropriate to the modern age of high computerization.

6 Super-Effective-Field Theory of Exotic Phase Transitions

In this section, we present a general scheme of the super-effective-field theory [25,56] to study exotic phase transitions. For this purpose we consider a general cluster shown in Fig.6. An effective field conjugate to a topological exotic order parameter is applied to the outer circles.

Then the "super-effective" Hamiltonian is given by

$$\tilde{\mathcal{H}} = \mathcal{H}_{\Omega} - \Lambda \sum Q_j \tag{6.1}$$

for any possible order parameter Q. A new aspect of this theory is to introduce a semi-local effective field Λ conjugate to the order parameter Q_j which is defined in the j−th schematic circle region in Fig.6. That is, the topological order Q_0 at the center emerges owing to the correlation through the window regions $D_1, D_2, \cdots, D_z$ and from the self-consistency condition

$$\langle Q_0 \rangle = \varepsilon_{0j} \langle Q_j \rangle \tag{6.2}$$

with some appropriate modular factor ε_{0j}. This extension of effective-field theory is very useful in studying exotic phase transitions such as spin glasses

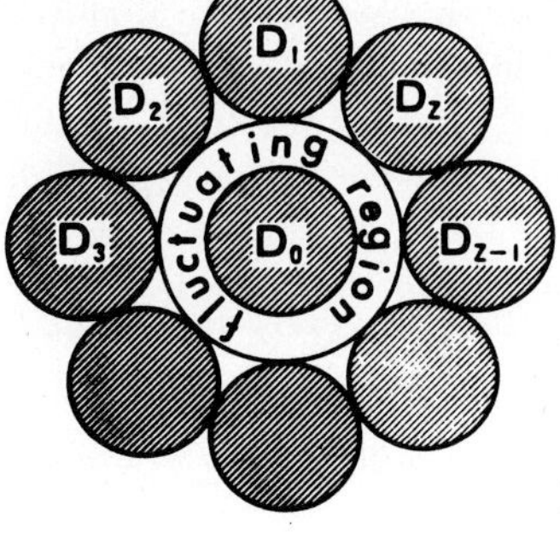

Fig.6. Super-effective-field cluster.

[25,57,58], chiral orders [25,59,60] and the KT-transition, as will be seen in the succeeding sections.

7 Spin Glasses

It is still a challenging problem to study spin glasses [61-68] in two and three dimensions for short-range interactions. Here we discuss the $\pm J$ Ising spin glass as an example of applications of the super-effective-field theory [56-58]. The order parameter of spin glasses is conveniently defined by the correlation between two real replicas, namely $\langle\langle S_0 S_0' \rangle\rangle_J$. Here the symbol $\langle\cdots\rangle$ denotes the average over the canonical ensemble and $\langle\cdots\rangle_J$ the average over the distribution of random interaction ($\pm J$). The effective Hamiltonian is given [18] by

$$\tilde{\mathcal{H}} = \mathcal{H}_\Omega^{(1)} + \mathcal{H}_\Omega^{(2)} - \Lambda \sum_{j \in \partial\Omega} S_j S_j', \tag{7.1}$$

where $\mathcal{H}_\Omega^{(j)}$ denotes the j-th replica cluster Hamiltonian. The self-consistency condition $\langle\langle S_0 S_0' \rangle\rangle_J = \langle\langle S_j S_j' \rangle\rangle_J$ gives an approximate critical point and spin-glass susceptibility which is equivalent to the negative nonlinear susceptibility [62]. Hatano and the present author [58] constructed several systematic super-effective-field approximations and applied the CAM to these results. Our conclusion is that there occurs a spin-glass transition in three dimensions with the critical point $T_{sg} \simeq 1.31$ and the critical exponent $\gamma \simeq 3.4$. These results are consistent with those obtained by other methods [63-65].

8 Chiral Orders in Spin Systems

8.1 Vector Chiral Orders in the XY Model and the Plane-Rotator Model

First we discuss here the vector chiral order in the plane-rotator model. The chirality is explained graphically [22-25] as in Fig.7. The simplest cluster for the super-effective-field theory of this problem is shown in Fig.8. This cluster corresponds to the general super-effective-field cluster shown in Fig.6. A chiral field Λ conjugate to the chiral order Q_{ijk} is applied at the outer triangular cell (ijk) of the cluster in Fig.8, and consequently the chiral order $\langle Q_0 \rangle$ at the center of the cluster is induced through the interaction inside the cluster. The self-consistency condition is then given by $\langle Q_0 \rangle = -\langle Q_{ijk} \rangle$. This procedure can be extended to large clusters. Thus, Kawashima and the present author [59] analyzed the chiral-order transition, applying the CAM theory to the results obtained by the above SEFT. Thus, we arrived at the conclusion that $T_c^* = 0.5096 \pm 0.0005$ and $\gamma \simeq 1.7 \pm 0.2$. These are consistent with the results obtained by Monte Carlo simulations.

Fig.7

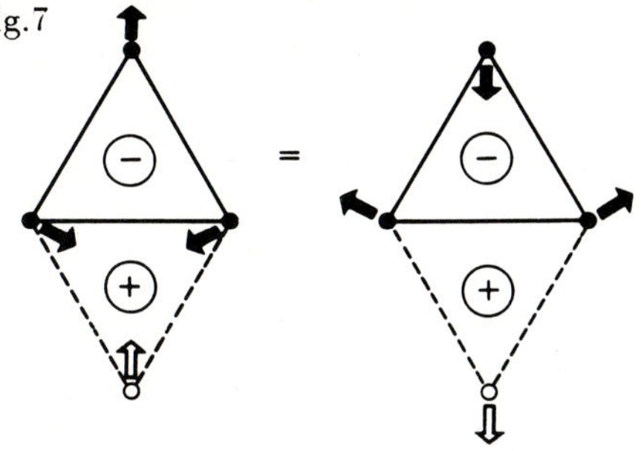

Fig.8

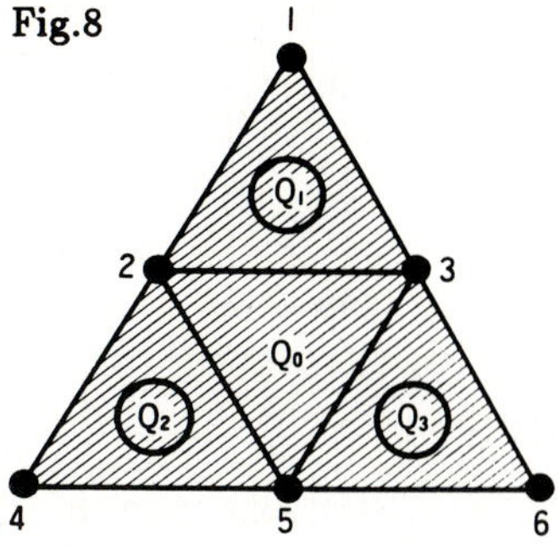

Fig.7 chiral orders in the plane rotator model [25].

Fig.8 A simple super-effective-field cluster for the chiral order [25].

Next, we discuss the quantum XY model. The chiral order for this system is defined [25,69] by the operator

$$Q_{ijk} = (\sigma_i \times \sigma_j + \sigma_j \times \sigma_k + \sigma_k \times \sigma_i)^z / 2\sqrt{3} \tag{8.1}$$

in terms of the Pauli operators σ_j^x, σ_j^y and σ_j^z for the triangular cell (ijk). The SEFT for the cluster shown in Fig.8 yields [25]the critical point $T_c \simeq 4.1 J/k_B$. This may be too large because of the present crudest approximation. The SEFT for larger clusters in the quantum XY model is much more complicated.

8.2 Scalar Chiral Order in Heisenberg Antiferromagnets

Quite recently the scalar chiral order defined by

$$\chi_{ijk} = \boldsymbol{S}_i \cdot (\boldsymbol{S}_j \times \boldsymbol{S}_k) \tag{8.2}$$

was introduced by Wen et al. [70] in connection with the high-T_c superconductivity. As is well known, this order parameter breaks the parity and the time- reversal symmetry.

Kawarabayashi and the present author [60] studied this scalar chiral order in the two-dimensional antiferromagnetic Heisenberg model with next nearest neighbour interaction using the systematic series of clusters shown in Fig.9. It is shown numerically [60] that the feed-back function (T) defined similarly to that in (3.7) is smaller than unity for any ratio of the nearest-neighbour and next-nearest-neighbour interactions. Therefore, using the SEFT, we arrive [60] at the conclusion that there occurs no chiral order in this system. This is also consistent with other calculations.

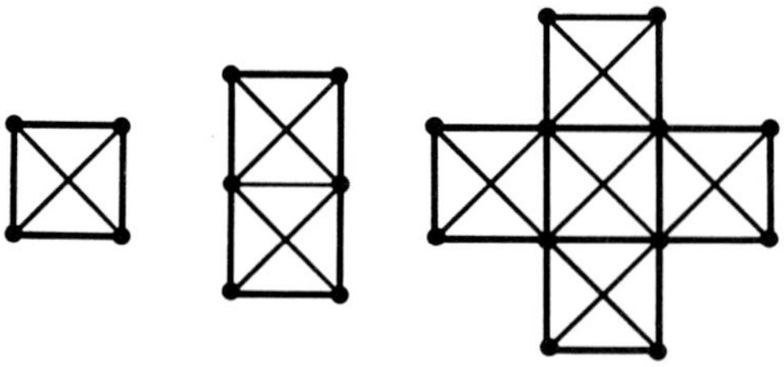

(a) (b) (c)

Fig.9 A systematic series of clusters for the SEFT.

8.3 Scalar Chiral Order in the Hubbard Model

It is more interesting to study the scalar chiral order in the t-J and Hubbard models with holes [71,72]. In the SEFT [72] based on the cluster (b) in Fig.9, we find that there occurs the scalar chiral-order transition in some extremely restricted region of the interaction-parameter space and hole concentration.

9 CAM Theory of the Kosterlitz-Thouless Transition

In the present section, we propose a CAM theory of the KT-transition. As is well known, this transition is caused by the condensation of vortices and there appears no long-range order even below the critical point in the two-dimensional planar model. In this sense, this transition may be called a "response diverging phase transition". In fact, the susceptibility $\chi_0(T)$ defined by $\chi_0(T) = \beta\langle M^2\rangle$ with $M = \sum S_j^x$ diverges for $T \leq T_c^*$. Thus, we make use of this "pseudo order parameter" M in order to construct an effective Hamiltonian of the form

$$\tilde{\mathcal{H}} = \mathcal{H}_\Omega - H_{\text{eff}} \sum_{j\epsilon\partial\Omega} S_j^x . \tag{9.1}$$

Then we construct systematic effective-field approximations in the form $\chi_0(T) = \chi_\Omega(T)/(1-\bar{\mathcal{H}}_\Omega(T)) \simeq \bar{\chi}(T_c)/(T-T_c)$ with the condition $\bar{\mathcal{H}}_\Omega(T_c) = 1$. Here $\bar{\chi}(T_c)$ or even $\chi_0(T_c)$ is to show the following essential singularity

$$\bar{\chi}(T_c)(\quad\text{or}\quad \chi_0(T_c)) \sim \exp\left(\frac{a}{\sqrt{T_c - T_c^*}}\right) . \tag{9.2}$$

Similarly we have

$$m_s(T) \simeq \bar{m}(T_c)(T_c - T)^{1/2} \qquad ; \; \bar{m}(T_c) \sim \exp\left(-\frac{b}{\sqrt{T_c - T_c^*}}\right) , \tag{9.3}$$

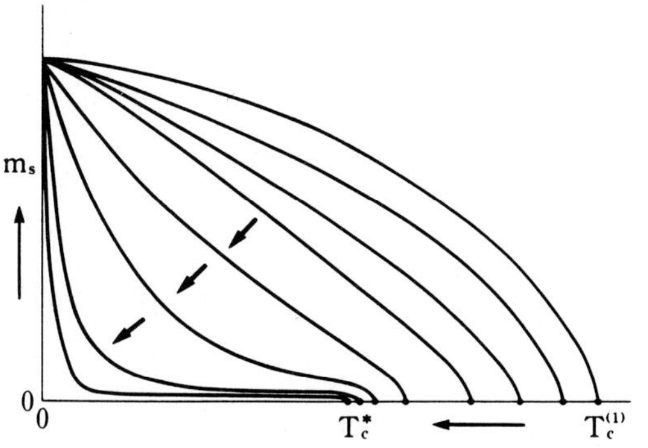

Fig.10 Schematic change of the pseudo spontaneous magnetization m_s obtained by the effective-field theory for the KT-transition using the pseudo order parameter M.

as the approximate critical point T_c approaches the true critical point T_c^*. This may lead to $m_s = 0$ below the critical point. A schematic change of the pseudo spontaneous magnetization m_s for increasing degree of approximations is shown in Fig.10. This scheme is quite remarkable in the sense that even highly topological orders such as the KT-transition can be studied using the CAM theory. More detailed explicit calculations will be reported elsewhere.

10 KT-Superconducting Transition in the 2d Negative - U Hubbard Model

The Hubbard model described by the Hamiltonian

$$\mathcal{H} = -t\sum_{ij,\sigma}(c_{i\sigma}^{\dagger}c_{j\sigma} + h.c.) + U\sum_{j}(n_{j\uparrow} - \frac{1}{2})(n_{j\downarrow} - \frac{1}{2}) - \mu\sum_{j\sigma} n_{j\sigma} \tag{10.1}$$

with fermion operators $c_{j\sigma}^{\dagger}, c_{j\sigma}$ and with $n_{j\sigma} = c_{j\sigma}^{\dagger}c_{j\sigma}$ has been studied extensively by many authors in connection with the high-T_c superconductivity. In particular, the negative-U case is also quite interesting , because there may occur [73-75] a KT-superconducting transition in this system.

Our scheme of the super-effective-field theory (SEFT) for this problem is to consider the following cluster effective Hamiltonian

$$\tilde{\mathcal{H}} = \mathcal{H}_{\Omega} - \Lambda\sum_{j\epsilon\partial\Omega}(c_{j\uparrow}^{\dagger}c_{j\downarrow}^{\dagger} + c_{j\downarrow}c_{j\uparrow}), \tag{10.2}$$

where $\mathcal{H}_{\Omega}$ denotes the original Hubbard model for the cluster Ω. The simplest effective-field theory is constructed by the following one-site effective Hamiltonian

$$\mathcal{H}_{\text{eff}} = U(n_\uparrow - \frac{1}{2})(n_\downarrow - \frac{1}{2}) - \mu(n_\uparrow + n_\downarrow) - \Lambda(c_\uparrow^\dagger c_\downarrow^\dagger + c_\downarrow c_\uparrow). \tag{10.3}$$

The partition function Z of this system is given by

$$Z = 2e^{\beta\mu - \beta U/4}(e^{\beta U/2} + \cosh[\beta(\mu^2 + \Lambda^2)^{1/2}]). \tag{10.4}$$

Consequently we obtain the electron number $\langle n \rangle$ and the Cooper pair density $\langle \hat{\Delta} \rangle = \langle c_\uparrow^\dagger c_\downarrow^\dagger + c_\downarrow c_\uparrow \rangle$ as

$$\langle n \rangle = 1 + (\mu/(\mu^2 + \Lambda^2)^{1/2}) f(\mu, \Lambda) \quad \text{and} \quad \langle \hat{\Delta} \rangle = (\Lambda/(\mu^2 + \Lambda^2)^{1/2}) f(\mu, \Lambda), \tag{10.5}$$

where

$$f(\mu, \Lambda) = \sinh[\beta(\mu^2 + \Lambda^2)^{1/2}]/(e^{\beta U/2} + \cosh[\beta(\mu^2 + \Lambda^2)^{1/2}]). \tag{10.6}$$

These expressions are useful in discussing effective-field treatments with the self-consistency condition that $\Lambda \sim -U \langle \hat{\Delta} \rangle$. Thus, we can determine the critical point T_c. It is the order of U in this crudest approximation. It is too high. In order to take into account the effect of transfer t and fluctuations, we have to study a two-site cluster and larger clusters systematically. These calculations will be reported in the near future elsewhere.

11 Summary and Future Problems

In the present paper, new trends of theories of phase transitions have been discussed with emphasis on the new approach to exotic phase transitions, namely the coherent-anomaly method. This approach together with the super-effective-field theory has been discovered by combining two basic concepts, namely Kubo's linear response and Fisher's finite-size scaling.

There remain many challenging problems on phase transitions, namely the high-T_c superconductivity, Anderson localization, critical phenomena of chaos and other exotic phase transitions in the gauge theory, nuclear physics and the universe.

Quite recently Zheng [76] pointed out the possibility to study, using the CAM, the Hausdorff dimension of Feigenbaum's limiting set of period doubling in the chaos problem. C.K. Hu [77] emphasized the usefulness of percolation representations of critical phenomena combined with the CAM theory. Many non-equilibrium phase transitions will be also studied [78] extensively using the CAM in future in connection with biological systems such as neural networks and brains. The present CAM and SEFT will be useful in analyzing such complex dynamical systems in the future.

References

[1] P.Weiss, J.Phys.Radium **6** (1907) 661.

[2] E.Ising, Z.Phys. **31** (1925) 253.

[3] L.Onsager, Phys.Rev. **65** (1944) 117.

[4] C.N.Yang, Phys.Rev. **85** (1952) 809.

[5] R.J.Baxter, Ann.Phys.(N.Y.) **70** (1972) 193.

[6] R.B.Griffiths, Phys.Rev.Lett. **24** (1970) 1479.

[7] L.P.Kadanoff, Proceedings of the Enrico Fermi Summer School, edited by M.S.Green (Academic Press, 1972).

[8] M.Suzuki, Prog.Theor.Phys. **51** (1974) 1992.

[9] A.M.Polyakov, JETP Lett. **12** (1970) 381.

[10] S.Howes, L.P.Kadanoff and M. den Nijs, Nucl.Phys. **B215** [FS7] (1983) 169.

[11] H.Au-Yang, B.M.McCoy, J.H.H. Perk, S.Tang and M.L.Yan, Phys.Letts. **A123** (1987) 219.

[12] L.D.Landau, Phys.Zhurn.Sowjetunion **11**(1937) 26, 545.

[13] M.E.Fisher, J.Math.Phys. **5** (1964) 944.

[14] B.Widom, J.Chem.Phys. **43** (1965) 3898.

[15] L.P.Kadanoff, Physics **2** (1966) 263.

[16] K.G.Wilson, Phys.Rev. **B4** (1971) 3174,3184.

[17] M.E.Fisher, J.Vac.Sci. & Tech. **10** (1973) 665.

[18] M.Suzuki, J.Phys.Soc.Jpn. **55** (1986) 4205.

[19] J.M.Kosterlitz and D.J.Thouless, J.Phys.C: Solid State Phys. **7** (1974) 1046.

[20] S.F.Edwards and P.W.Anderson, J.Phys.F ; Metal Phys. **5** (1975) 965.

[21] G.Toulouse, Commun.Phys. **2** (1977) 115.

[22] J.Villain, J.Phys. **C10** : 1717 and 4793 (1977).

[23] S.Miyashita and H.Shiba, J.Phys.Soc.Jpn. **53** (1984) 1145.

[24] D.H.Lee, J.D.Joannopoulos, J.W.Negele and D.P.Landau, Phys.Rev.Lett. **52** (1984) 433 ; Phys.Rev. **33** (1986) 450.

[25] M.Suzuki, J.Phys.Soc.Jpn. **57** (1988) 2310.

[26] M.Suzuki, Prog.Theor.Phys. **46** (1971) 1337.

[27] M.Suzuki, Prog.Theor.Phys. **56** (1976) 1454.

[28] M.Suzuki, S.Miyashita and A.Kuroda, Prog.Theor.Phys. **58** (1977) 1377.

[29] T.D.Lee and C.N.Yang, Phys.Rev. **87** (1952) 410.

[30] M.Suzuki, J.Phys.Soc.Jpn. **60** (1991) 445.

[31] C.N.Yang, Rev.Mod.Phys. **34** (1962) 694.

[32] R.Kubo, J.Phys.Soc.Jpn. **12** (1957) 570.

[33] H.A.Bethe, Proc.Roy.soc. **A150** (1935) 552.

[34] M.Suzuki, M.Katori and X.Hu, J.Phys.Soc.Jpn. **56** (1987) 3092.

[35] M.Katori and M.Suzuki, J.Phys.Soc.Jpn. **25** (1987) 3113.

[36] X.Hu, M.Katori and M.Suzuki, J.Phys.Soc.Jpn. **56** (1987) 3865.

[37] X.Hu and M.Suzuki, J.Phys.Soc.Jpn. **57** (1988) 791.

[38] M.Katori and M.Suzuki, J.Phys.Soc.Jpn. **57** (1988) 807.

[39] M.Suzuki, in **Thermal Field Theories and Their Applications**, eds.H.Ezawa, T.Arimitsu and Y.Hashimoto, Elsevier Science Pub. (1991) and references cited therein.

[40] M.Suzuki, in **New Trends in Magnetism**, ed. by M.D.Coutino-Filho and S.M. Rezende (World Scientific, Singapore, 1990).

[41] K.Minami, Y.Nonomura, M.Katori and M.Suzuki, Physica A (1991).

[42] A.Lipowski, Physica A **173** (1991) 293.

[43] T.Oguchi, J.Phys.Soc.Jpn. **30** (1971) 988, ibid **50** (1983) 3101.

[44] T.Oguchi and I.Ono, J.Phys.Soc.Jpn. **21** (1966) 2178.

[45] T.Oguchi and H.Kitatani, J.Phys.Soc.Jpn. **57** (1988) 3973.

[46] J.O.Indeku, A.Maritan and A.L.Stella, J.Phys.A (Math. Gen.) **15** (1982) L291.

[47] Y.Nonomura and M.Suzuki, submitted to J. Phys. A.

[48] B.M.McCoy, Phys.Rev. **173** (1968) 531.

[49] R.J.Glauber, J.Math.Phys. **4** (1963) 294.

[50] M.Suzuki and R.Kubo, J.Phys.Soc.Jpn. **24** (1968) 51.

[51] N.Ito, M.Taiji and M.Suzuki, J.Phys.Soc.Jpn. **56** (1987) 4218.

[52] H.Yahata and M.Suzuki, J.Phys.Soc.Jpn. **27** (1969) 1421.

[53] M.Suzuki, J.Phys.Soc.Jpn. **56** (1987) 4221.

[54] M.Suzuki, J.Phys.Soc.Jpn. **57** (1988) 1.

[55] T.Oguchi and H.Kitatani, J.Phys.Soc.Jpn. **57** 81988) 3973.

[56] M.Suzuki, J.Stat.Phys. **53** (1988) 483.

[57] M.Suzuki, J.de Phys. (Paris) Colloque **8** (1988) 1519.

[58] N.Hatano and M.Suzuki, J.Stat.Phys. **63** (1991) 25 and in press.

[59] N.Kawashima and M.Suzuki, J.Phys.Soc.Jpn. **58** (1988) 3123.

[60] T.Kawarabayashi and M.Suzuki, J.Phys.Soc.Jpn. **60** (1991) 825.

[61] S.F.Edwards and P.W.Anderson, J.Phys.F ; Metal Phys. **5** (1975) 965.

[62] M.Suzuki, Prog.Theor.Phys. **58** (1977) 1151.

[63] A.T.Ogielski and I.Morgenstern, Phys.Rev.Lett. **54** (1985) 928.

[64] R.N.Bhatt and A.P.Young, Phys.Rev.Lett. **54** (1985) 924.

[65] R.R.P.Singh and S.Chakravarty, Phys.Rev.Lett. **57** (1986) 245.

[66] D.Sherrington and S.Kirkpatrick, Phys.Rev.Lett. **35** (1975) 1972.

[67] K.Binder and A.P.Young, Rev.Mod.Phys. **58** (1986) 801.

[68] D.Chowdhury, **Spin Glasses and Other Frustrated Systems**, World Sci. 1986.

[69] S.Fujiki and D.D.Betts, Can.J.Phys. **64** (1986) 876 ; ibid **65** (1987) 76. S.Fujiki and D.D.Betts, Prog.Theor.Phys.Suppl. **No 87** (1986) 268.

[70] W.G.Wen, F.Wilczek and A.Zee, Phys.Rev. **B39** (1989) 11413. Yi-H Chen, F.Wilczek, E.Witten and B.I.Halperin, Int.J.Mod.Phys. **B3** (1989) 1001.

[71] O.Narikiyo, K.Kuboki and H.Fukuyama, J.Phys.Soc.Jpn. **59** (1990) 2443.

[72] Y.Nonomura and M.Suzuki, in preparation.

[73] J.E.Hirsch and S.Tang, Phys.Rev.Lett. **62** (1989) 591.

[74] R.T.Scaletter, E.Y.Loh, J.E.Gubernatis, A.Moreo, S.R.White, D.J.Scalapino, R.L.Sugar and E.Dagotto, Phys.Rev.Lett. **62** (1989) 1407.

[75] A.Moreo and D.J.Scalapino, Phys. Rev. Lett. **66** (1991) 946.

[76] Wei-M.Zheng, Phys.Lett. **143A** (1990) 362.

[77] C.K.Hu, Proceedings of the Fourth Asia Pacific Physics Conference, Seoul, Korea, 13-17, August (1990), vol.1 p.21 (1991) World Scientific, edited by S.H.Ahn, S.H.Choh, Il-T.Cheon and C.Lee.

[78] B.Schmittmann, Int.J.Mod.Phys. **B4** (1990) 2269.

Nuclear Fusion, Its Physics and Technology

B.B. Kadomtsev

I.V. Kurchatov Institute of Atomic Energy,
46 Kurchatov St., 123182, Moscow, USSR

Abstract. The main lines of progress in physics and technology in the field of controlled nuclear fusion research are discussed in this paper.

1. Introduction

One of the most striking features of our century is the broad penetration of advanced technologies into all areas of social life. This concerns both civil and military aspects of the modern world: new technologies have changed the style of life and even the face of modern society. Many of these new technologies emerged from basic science and some of them were born from theoretical physics. We can claim that the theoretical physics developed by prominent physicists of our century, including Professor Y. Nishina, has provided a basis for the most important inventions which have changed the modern world.

Controlled nuclear fusion was suggested initially as a purely theoretical idea. In our country this was done by Professor I.E. Tamm and Professor A.D. Sakharov, Nobel prize winners. Scientific activity in this field has passed along the very long road of persistent experimental and theoretical effort and is now reaching the stage where the design and then construction of the Experimental Thermonuclear Reactor can be begun.

2. Plasma Heating and Confinement

To ignite the nuclear fusion reaction, the mixture of the hydrogen isotopes deuterium and tritium has to be heated and transferred in the state of an ionized gas with the temperature above several keV (1 keV is equivalent to about ten million degrees Celsius). The plasma matter has to be sustained during some period of time to provide the fusion fuel burn. The main issues which have to be assessed are plasma heating and plasma confinement.

There are two basically different approaches to plasma confinement: either the use of a magnetic field for insulation of the rare plasma from the cold chamber walls or the use of inertially delayed expansion of the hot and very dense pellets. Both approaches are now being investigated with different heating schemes for different concepts of confinement.

Springer Proceedings in Physics, Volume 57
Evolutionary Trends in the Physical Sciences Eds.: M. Suzuki and R. Kubo

For magnetic confinement the so-called tokamak concept is the most advanced [1]. At present the tokamaks are close to the so-called break-even point, when the fusion power is equal to the heating power needed to sustain the plasma at steady state. In conjunction with the tokamak concept many advanced fusion technologies have been developed. The tokamak concept was chosen for the design of the Experimental Thermonuclear Reactor.

For inertial confinement the laser fusion approach is the most advanced now. The laser fusion approach uses pellets of solid D-T fuel which have to be first compressed to very high density and then heated to the temperature required for ignition.

3. Tokamak Physics

The idea of magnetic confinement in tokamak geometry relies on the fact that a magnetic field can strongly affect the motion of plasma electrons and ions. If the plasma is not dense, macroscopic electric currents can be easily generated in it. As a result, a closed confinement configuration originates which is toroidal, i.e. doughnut shaped. The strong toroidal magnets' field is added to provide plasma stability. Figure 1 shows a family of nested toroidal surfaces, produced by the magnetic field lines. To reach the wall the individual plasma particles (electrons and ions) have to cross magnetic surfaces. However, such a motion is strongly slowed down in the strong magnetic field.

3.1 Plasma Confinement

The theory which describes the cross-field transport due to interparticle collisions was suggested firstly by A.A. Galeev and R.S. Sagdeev [2] and developed in detail by many theorists. This theory takes into account the details of particle trajectories in the magnetic field. It is called the "neoclassical theory". Many years of experimental research have shown that the real plasma behaviour can

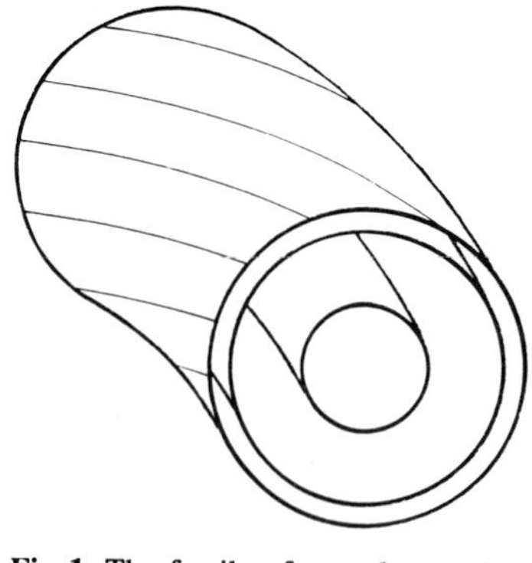

Fig. 1. The family of nested smooth toroidal magnetic surfaces present in tokamak geometry when the ideal axial symmetry is established

be very different from the theoretical predictions. This discrepancy is not very large for ions, but for electrons the real cross-field transport differs by a factor of hundreds from that predicted. It was shown theoretically that the real geometry is quite different from Fig. 1. In real experiments the axial magnetic symmetry in tokamaks is broken. A very complicated set of slightly distorted toroidal layers appears instead of smooth magnetic surfaces. Similar to the other physical patterns like domains in ferromagnetics, the set of vortices in rotating He II, and the set of fluxons in type-II superconductors, the smooth magnetic surfaces in tokamaks are disintegrated into a mixture of filaments. It is better to say that some kind of current filamentation takes place in the tokamak plasma. The spontaneous broken symmetry in tokamaks differs from the other examples because it has the nature of a highly nonequilibrium phenomenon of self-organization.

A single unit of such a pattern looks like a ring of magnetic tubes aligned with each other (Fig. 2).

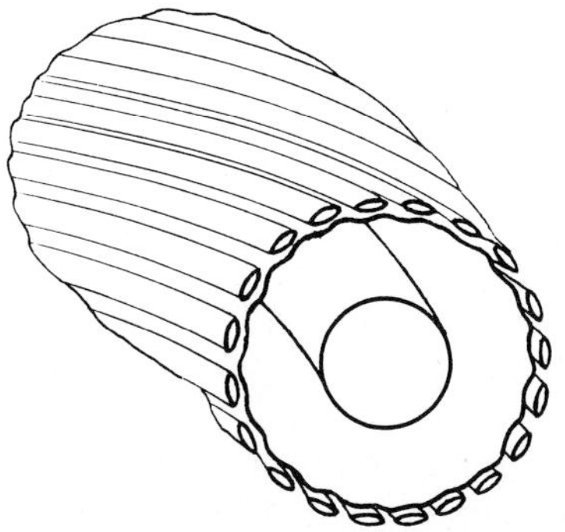

Fig. 2. The real tokamak magnetic surfaces are destroyed by themselves: the spontaneously broken symmetry looks like a chain of filaments. The cross-section size of each filament is much less then the mean value of the ion Larmor radius but is larger than the mean electron Larmor radius. Rotating in the direction of ion diamagnetic drift the filament pattern is pumped up by ions

The tokamak magnetic field line geometry looks like a mixture of nested toroidal magnetic surfaces with island-chain patterns immersed somewhere between them. Each island-chain rotates in the direction of the ion diamagnetic rotation and is slightly pumped by ions. Many island-chains fill all the torus cross-section and produce both ion and electron transport. The physics of the magnetic noise pumping is illustrated in Fig. 3.

Let us assume that the magnetic islands have a small deviation from the initial magnetic surface position (Fig. 3). In this case the ions can be "blown through" the island-chain in the direction of the lower magnetic field, giving away a part of their energy and transferring it to the magnetic noise. This noise pumping is responsible for the plasma microturbulence which in turn produces enhanced plasma transport. It was found that ion and electron motions are marginally stochastic so that the plasma transport is very sensitive to the internal and external conditions. The transport becomes self-controlled, having feedback coupling

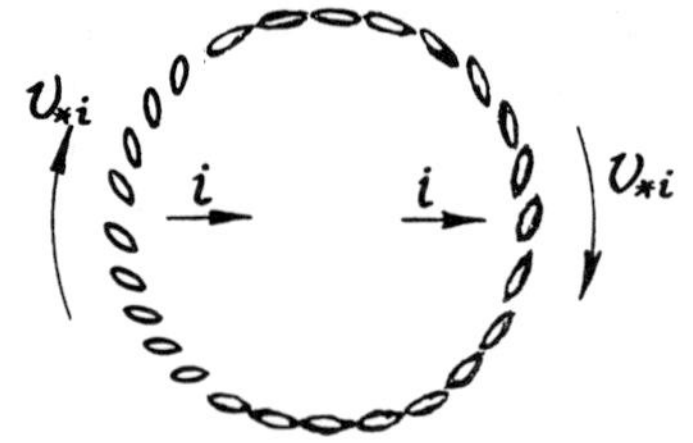

Fig. 3. If the island-chain pattern rotates in the direction of the ion diamagnetic rotation and the islands are inclined with respect to the rational surface (where the islands were initiated), as shown in the figure, the ions can be shifted in the direction of the magnetic field decrease. As the ion adiabatic invariant is conserved, each ion will lose part of its energy, transferring it into island-chain pump up

with the profiles of the plasma pressure and current density as functions of the minor radius (the distance from the magnetic axis). This phenomenon observed experimentally in many tokamaks is called "profile consistency" [2].

The plasma transport in tokamaks is very sensitive to the boundary conditions. Edge plasma is turbulent. The level of this turbulence can be lowered with the help of plasma differential rotation near the wall. Such a rotation can build up by itself and give rise to an improved plasma confinement. In this mode the plasma confinement capability is sufficient to proceed to the experimental tokamak-reactor design.

3.2 Control of Impurities

The fusion plasma has to be clean. This means that some tools have to be present in the tokamak-reactor to purge the plasma from impurities. This issue can be resolved with the help of the magnetic divertor configuration shown in Fig. 4.

The plasma boundary in this configuration is defined by the transition between closed, nested magnetic surfaces and the open magnetic field lines, which eventually intersect target plates away from the main plasma.

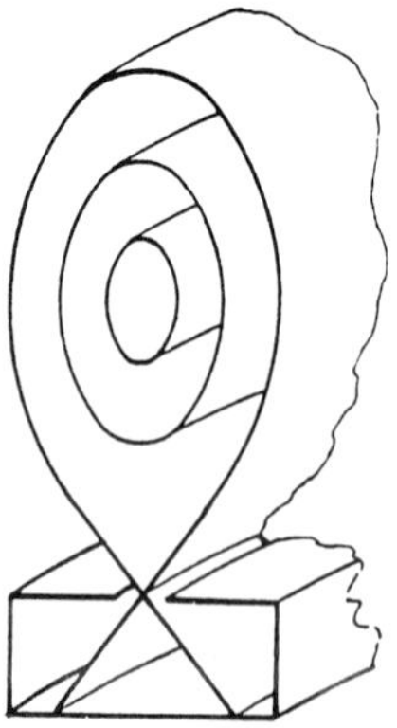

Fig. 4. Tokamak magnetic configuration with the single-null poloidal divertor

Most advanced tokamaks, like the JT-60, JET and DIII-D, have such a configuration. The divertor configuration was also accepted for ITER – International Thermonuclear Experimental Reactor – being designed at present by the EC, Japan, USSR and USA under the auspices of the IAEA.

3.3 Current Drive and Heating

The future thermonuclear reactor has to be a steady-state machine. Steady-state operation in a tokamak can be established with the help of noninductive current drive schemes. They rely on auxiliary heating capabilities. To increase the plasma temperature an auxiliary heating power has to be introduced into the tokamak. This power can be used for momentum transfer to electrons carrying the plasma current. There are several schemes for plasma heating and current drive: injection of a beam of high energy neutral particles (NBI), electromagnetic waves in different frequency ranges, such as ion cyclotron heating (ICRH), electron cyclotron resonance heating (ECRH) and lower hybrid resonance (LH) heating. All of them are used both for plasma heating and plasma current drive.

Record ion temperatures of about 30 keV, i.e. about twice the value needed for ignition, have been achieved in TFTR [4] and JET [5]. In JT-60 [6], a current of 2 MA has been driven noninductively by lower hybrid waves. Current drive for about one hour has been achieved on the superconducting tokamak TRIAM-1M [7].

The current drive in tokamaks is accompanied by the so-called "bootstrap" current predicted by neoclassical theory [2]. It is proportional to the plasma pressure gradient and therefore a large bootstrap current is expected to be generated in future tokamak-reactors. The "bootstrap" current was observed in several tokamaks (for example, in TFTR and JET). Recently, in JT-60 the "bootstrap" current was estimated to be up to 80% of the total current (0.5 MA) during heating by 20 MW NBI.

3.4 Status of Tokamak Physics

The main figure of merit used to assess the approach to ignition is the ratio Q of the fusion power released by D-T reaction to the power lost from the plasma. At ignition-relevant temperatures this ratio is proportional to the triple product $n\tau T$, where n is the plasma density, T its temperature and τ the energy confinement time (i.e. the ratio of plasma energy to the heating power for the steady-state plasma).

At present the Q-value projected to the D-T mixture is close to unity for JET and is close to 0.5 for TFTR. This means that the "break-even" point has almost been reached. An ion temperature of about 30 keV has been achieved in TFTR and JET. The physics data base is sufficient for design and then construction of the thermonuclear reactor.

4. Tokamak Technology

The most advanced fusion technologies have been developed for tokamaks. Many of them can be used for alternative approaches with the magnetic confinement of plasma.

4.1 Magnets

Superconducting magnets will be the most expensive components required for a fusion reactor. So far, superconducting magnets have been built for laboratory size tokamaks: T-7, TRIAM, TORUS-SUPRA, T-15. Magnets of the size, field, current and reliability required for a reactor have to be fabricated and tested.

4.2 Auxiliary Heating and Current Drive

The heating tools for tokamak plasma have been developed and tested on the existing machines. They have to be further developed and enlarged for reactor application.

4.3 Nuclear Technology

The first fusion reactor will use the deuterium-tritium fuel. The fusion reactor fuel cycle system has to provide all tritium handling, coupled both with the toroidal chamber and with the tritium breeding blanket.

This technology has to be developed on the basis of existing tritium technologies and to be tested for fusion reactor applications.

4.4 Materials

The plasma-facing materials will be a long-term issue of fusion technology. The first phase is the developing and testing of plasma-facing materials which are acceptable from the plasma physics point of view. In the second phase more advanced materials have to be developed and tested. These materials have to meet more severe conditions in the future fusion reactors. Low activated materials are anticipated to be used in future reactors.

4.5 Safety and Environment

The safety and environmental impact of fusion reactors are important issues which are being studied intensively [8]. Fusion reactors will have substantial advantages over fission reactors with respect to the consequences of severe accidents and the magnitude of radioactive-waste burdens. The maximum critical dose at the site boundary for a severe fusion accident is two to three orders of magnitude less than that for a severe fission accident.

5. ITER as a Next Step

The ITER device is presently being designed by four parties (EC, Japan, USSR and USA) under the auspices of the IAEA. The device is for controlled ignition and extended burn of D-T plasmas, as well as for demonstration and integrated testing of components required to utilize fusion power for practical purposes [9]. The ITER conceptual design results from extensive research and development over several decades. The design, construction and operation of a tokamak within the ITER definition will provide numerous technological challenges that will have to be met in order for a fusion reactor to be realized. Credible solutions to all of these challenging tasks have been proposed. These solutions will be tested in the frame of an engineering and technological R&D programme.

6. Inertial Confinement Fusion

Inertial Confinement Fusion (ICF) represents an alternative line of nuclear fusion power assessment.

6.1 ICF Physics

Inertial confinement fusion accepts the implosion of pellets as the basic physics approach. Pellet implosion experiments can be accomplished with the help of different drivers: lasers, relativistic electrons and light ion beams, heavy ion beams. The physics of beam-target interaction is slightly different for different drivers so that targets have to be adjusted for each particular driver.

Two approaches are being investigated:

- direct drive approach, where a number of beams of photons or ions are made to converge on the target;
- indirect drive approach, where either the laser or the ion beam is first converted to soft X-rays that isotropically fill a metallic cavity (Hohlraum). The fuel pellet, placed in this cavity, is then symmetrically compressed and ignited.

Currently, the Nd glass laser is the most flexible and advanced driver. The most advanced facilities have been constructed in the USA, Japan and France. The Nova facility in the USA is capable of providing pulse energy in the range of 120 kJ in the near infra-red and about 50 kJ in blue light.

Laser fusion experiments have demonstrated D-T compressions up to 600 times the solid density. In some experiments, 10^{13} fusion neutrons have been produced, representing a fusion energy of 0.2% of the 10 kJ laser pulse.

The results of investigation of the implosion process with high power lasers have made it possible to design an IFC experiment with the aim of achieving ignition and break-even.

6.2 ICF Technology

ICF technology includes the very big and expensive driver facilities. Different drivers have been developed:

- Nd glass laser
- krypton fluoride excimer laser
- iodine laser
- light ion beams
- heavy ion beams
- relativistic electron beams

Many of them are being used for ICF experiments. Steady progress has been made in relevant areas, for example, in pellet fabrication, the design of the reactor chamber, the diagnostics and simulation programmes.

7. Conclusion

Fusion research programmes worldwide have made steady progress during the last decade. The tokamaks JET in Europe and TFTR in the USA have closely approached conditions equivalent to energy break-even, at which point as much fusion power is produced as is required to maintain the steady-state plasma. In inertial fusion, laboratory experiments using laser drivers are approaching 1% gain. Both directions are ready to proceed further.

The next-step device, which is at present at the stage of ITER conceptual design completion, aims at fully confirming the scientific feasibility and at addressing the technological feasibility of fusion as a potentially safe and environmentally acceptable, practically inexhaustible source of energy.

References

1. B.B. Kadomtsev, F.S. Troyon, M.L. Watkins: Tokamaks, Nucl. Fusion **30**, 1675 (1990)
2. A.A. Galeev, R.S. Sagdeev: In *Reviews of Plasma Physics*, Vol. 7, ed. by M. Leontovich (Consultants Bureau, New York 1979)
3. B. Coppy: Neoclassical Transport and the "Principle of Profile Consistency". Comments Plasma Phys. Contr. Fusion **5**, 2621–269 (1980)
4. M.G. Bell, V. Arunasalam, C.W. Barnes, et al.: In *Plasma Physics and Controlled Nuclear Fusion Research 1988* (Proc. 12th Int. Conf., Nice 1988), Vol. 1 (IAEA, Vienna 1989) p. 27
5. JET Team: In *Plasma Physics and Controlled Nuclear Fusion Research 1988* (Proc. 12th Int. Conf., Nice 1988), Vol. 1 (IAEA, Vienna 1989) p. 41
6. JT-60 Team: In *Plasma Physics and Controlled Nuclear Fusion Research 1988* (Proc. 12th Int. Conf., Nice 1988), Vol. 1 (IAEA, Vienna 1989) p. 67
7. S. Itoh, N. Hiraki, V. Nakamura, et al.: In 13th Int. Conf. on Plasma Physics and Controlled Nuclear Fusion Research, Washington, 1–6 October 1990, paper IAEA-CN-53/E-2-6
8. Status Report on Controlled Thermonuclear Fusion. Nucl. Fusion **30**, 1641 (1990)
9. International Atomic Energy Agency, ITER Concept Definition, Vols. 1 and 2, ITER Documentation Series, No. 3 (IAEA, Vienna 1989)
10. S. Nakai: Inertial Confinement. Nucl. Fusion **30**, 1863 (1990)

Cold Fusion – Does It Have a Future?

J. Schwinger

Department of Physics, University of California,
Los Angeles, CA 90024, USA

Abstract. The case against the reality of cold fusion is outlined. It is based on preconceptions inherited from experience with hot fusion. That cold fusion refers to a different regime is emphasized. The new regime is characterized by intermittency in the production of excess heat, tritium, and neutrons. A scenario is sketched, based on the hypothesis that small segments of the lattice can absorb released nuclear energy.

Minasama. Ladies and Gentlemen.

A totally unexpected phenomenon has been discovered in a certain field of science. It could have significant implications for the future of mankind, and especially for the Japanese. The overwhelming reaction of the experts in the field is rejection, based on the absence of other effects that are considered to be necessary companions of this new phenomenon. To quote one expert: "We know a lot about what happens . . . . We no longer have the latitude to say 'Well, some strange event occurred and generated those things.'" Nevertheless, this new possibility seems to have enough validity that one skeptic said: "It's hard to believe it. But there seems to be something to this." And, he went on to say: "It should not be necessary, however, to understand the mechanism before embracing the concept. If a proven track record can be established . . . you have to believe it."

To which scientific field does all this refer? In view of the title of my lecture, the question may seem surprising. In fact, the subject is *seismology*. The new phenomenon is the occurrence of electromagnetic effects just prior to the onset of an earthquake. The most striking event happened on 17 October, 1989. The apparatus of a team of radio detection specialists, which was situated in the Santa Cruz mountains of California, received an unprecedented blast of radio power. The strong signal continued for several hours, and then stopped, to be followed by the Loma Prieta earthquake that, last year, wreaked severe damage in the San Francisco area. Another kind of measurement seeks changes in electrical resistivity for ground currents. Scientists at Athens, Greece, have established a track record of 75% success in predicting earthquakes.

Of course — apart from the specific words of the quotations — all that I said before also applies to the phenomenon of cold fusion.

Springer Proceedings in Physics, Volume 57
Evolutionary Trends in the Physical Sciences Eds.: M. Suzuki and R. Kubo

It is astonishing that there was an early precursor of the claim to have achieved cold fusion. Dated at the beginning of the Showa era, the German title of the paper is translated as "On the transformation of hydrogen into helium." At that time, neither the existence of the heavier isotopes of hydrogen, nor of the lighter isotope of helium, was recognized. If, indeed, they did produce helium, was it ^{4}He, or was it ^{3}He? Incidentally, at just that time, Nishina Yoshio was at Niels Bohr's institute in Copenhagen. One can only wonder how he reacted to the bizarre claim.

On 23 March, 1989, the University of Utah, at Salt Lake City, threw a press party. Its purpose was to establish priority for patents on a new source of energy. The impetus was supplied by what seemed to be a rival group, down the road at Provo, Utah. The patent lawyers needn't have worried. The Provo people were investigating a very weak source of neutrons, which is only of academic interest. But, science that is filtered by patent attorneys is no longer science. Isn't it possible to establish a track record without reference to the initial claimants?

The National Cold Fusion Institute has provided a clearing house for reports that bear on the reality of cold fusion. As of August, 1990, 78 other groups, from all over the globe, have reported positive evidence, as conveyed by the detection of one or more of these indicators: excess heat, tritium, neutrons, γ-rays, ^{3}He. The standard response to such a list is: "Yes, but what about the much larger number of failures?" Does anyone really think that scientific judgment is like an election, in which the majority carries the day?

The characteristics that seem to be common to all successful cold fusion experiments are: (1) Intermittency — the production of heat, of tritium, of neutrons, comes in bursts, switching on and off at random. (2) Irreproducibility — seemingly identical cells vary widely in their ability to "turn on." It may not be too much of an exaggeration to say that, early in April, 1989, everyone — including those who, like myself, had to look up the meaning of enthalpy — had thrown together an electrolysis apparatus and was waiting for dividends. After a few weeks, with no reward, they quit in disgust, and denounced it all as incompetence, or fraud. Their votes are irrelevant.

Reproducibility is often cited as a canon of science. And so it is, in established areas. But, early in a study of a new phenomenon that involves an ill-understood macroscopic control of a microscopic mechanism, irreproducibility is not unknown. That was so at the onset of microchip studies. It also appeared in the initial phase of the discovery of high temperature superconductivity, which, by the way, is a prime example of "embracing the concept" without having "to understand the mechanism."

What is it about cold fusion that seems to enrage a substantial number of physicists? The people who have spent a lot of money on hot fusion would doubtless echo: "We know a lot about what happens . . . . We no longer have the latitude to say 'Well, some strange event occurred and generated these things.'" To be specific, this is how their preconceptions work: (1) In hot fusion, the union of two deuterons, to form ^{3}He and a neutron, proceeds at about the same rate as the formation of a triton and a proton. But the emission of neutrons from palladium electrodes immersed in heavy water occurs at a rate around the insignificant background level. Conclusion: No neutrons — no cold fusion. (2) The two cited reactions are the only important ones in hot fusion. There is no independent source of excess heat. Conclusion: Incompetence. (3) Given the essential absence of neutrons, what of the claims for substantial tritium production? Conclusion: Fraud. (4) At the low energy of cold fusion, the penetrability of the Coulomb barrier is so overwhelmingly small that nothing could possibly happen. Conclusion: Stupidity.

The next items of the hot fusioneer's creed are responses to suggested cold fusion mechanisms: (5) Very soon after 23 March, 1989, it was proposed that excess heat is produced by the formation of ground state ^{4}He in the DD fusion process. Response: Where is the accompanying γ-ray of roughly twenty million electron volts? (6) Then came the recognition that excess heat might be dominated by the HD, rather than the DD reaction. Heavy water unavoidably contains some fraction of a percent of light water. The fusion of a proton with a deuteron produces ^{3}He. Response: Where is the accompanying γ-ray of roughly five million electron volts? (7) The HD reaction is a source of heat and of ^{3}He, but not of neutrons or tritium. The latter must come from the DD reaction. What happens if two fusing deuterons populate, not the ground state, but the first excited state of ^{4}He? That excited state is unstable against decay into a triton and a proton. It is *stable*, however, for decay into a neutron and ^{3}He. Here, then, is a mechanism to account for the great disparity between neutron and triton production — the ratio is about one in a hundred million — that seems to be characteristic of cold fusion. Response: Where is the accompanying γ-ray of about four million electron volts?

So stands the indictment of cold fusion. The defence is simply stated: The circumstances of cold fusion are not those of hot fusion.

It is standard operating procedure, in hot fusion work, to represent the reaction rate as the product of two factors: the barrier penetration probability, which involves only the Coulomb repulsion; and, the intrinsic reaction rate, which is dominated by nuclear forces. But, at the very low energy of cold fusion, one is dealing, essentially, with a single wavefunction, which does not permit such factorization. The effect of Coulomb repulsion cannot be completely isolated from the effect of the strongly attractive nuclear forces. This is a whole new ballgame. It is, so to speak, a sumo tournament restricted to the maku-no-uchi, indeed, to the yokozuna.

The wavefunctions for a low energy proton and deuteron, and for a low energy pair of deuterons, are effectively dominated by zero relative angular momentum. They are states of even orbital parity. The intrinsic parities of all relevant particles — neutron, proton, deuteron, triton, ^{3}He, ground state, and first excited state of ^{4}He — are also positive. So, the normally dominant process of electric dipole radiation is forbidden; it requires a parity change.

If the γ-rays demanded by the hot fusioneers are greatly suppressed, what agency does carry off the excess energy in the various reactions? One must look for something that is characteristic of cold fusion, something that does not exist in the plasma regime of hot fusion. The obvious answer is: the lattice in which the deuterium is confined.

Imagine, then, that a small, but macroscopic piece of the lattice absorbs the excess energy of the HD or DD reaction. Please — I beg of you — do not rise in high dudgeon to protest that this is impossible because of the great disparity between atomic and nuclear energy scales. That is a primitive reaction to what may be a very sophisticated mechanism. And do not forget the failure of theory to predict, and then to account for the phenomenon of high temperature superconductivity. I advance the idea of the lattice playing a vital role as a *hypothesis*. Past experience dictates that I remind you that a hypothesis is not something to be proved mathematically. Rather, it is a basis for correlating data and for proposing new tests, which, by their success or failure, support or discredit the validity of the hypothesis. It is the essence of the scientific method.

Intermittency is the hallmark of cold fusion. It incorporates irreproducibility as a circumstance in which the time intervals between bursts significantly exceed the duration of the observations. Intermittency is the ultimate rebuttal to charges of fraud in tritium production. Externally introduced tritium maintains an essentially constant counting rate. There is no resemblance to the switching on and off of the observed bursts. Does the lattice hypothesis have a natural explanation for intermittency?

One needs information about the lattice structure of deuterided palladium. The experts say that "We know a lot . . . ", but that knowledge does not include what happens in the important regime of heavy deuteron loading. There is, however, a theoretical suggestion that, in the circumstance of heavy loading, a pair of new equilibrium sites comes into existence within each lattice cell. The equilibrium separation for that pair is significantly smaller than any other such distance in the cell.

It would seem that a close approach to saturation loading is required for effective fusion to take place. But, surely, the loading of deuterium into the palladium lattice does not occur with perfect spatial uniformity. There are fluctuations. It may happen that a microscopically large — if macroscopically small — region attains a state of such lattice uniformity that it can function collectively in absorbing the excess nuclear energy that is released in an act of fusion. And that energy can initiate a chain reaction as the vibrations of the excited ions bring them into closer proximity. So begins a burst. In the course of time, the increasing number of vacancies in the lattice will bring about a shut-down of the burst. The start-up of the next burst is an independent affair.

This scenario raises an interesting question: Would the efficacy of room temperature cold fusion be enhanced significantly by further lowering of the ambient temperature? Lower temperature would presumably decrease somewhat the probability of the initial fusion. But, it should increase the probability of forming and maintaining the lattice structure against the destructive onslaughts of thermal agitation. Experiment must supply the answer.

I find it both amusing and tragic that the members of a panel, investigating the charge of fraud in tritium production by cold fusion, dismissed the charge as "unlikely" and "much less probable than that of inadvertent contamination or other unexplained factors in the measurement." That the "unexplained factors" might be the reality of cold fusion is not admitted. Why? Because "critics questioned the results, saying that the tritium was not accompanied by other fusion byproducts . . . ." It is the old story. If a significant flux of neutrons is not observed, there cannot be any tritium, even though one finds tritium with a signature that differentiates it both from external and internal contamination.

The pressure for conformity is enormous. I have experienced it in editors' rejections of submitted papers, based on the venomous criticism of anonymous referees. The replacement of impartial reviewing by censorship will be the death of science.

Does cold fusion have a future? I have little hope for it in Europe and the United States — the West. It is to the East, and, specifically, to Nihon, that I turn. The willingness that the Japanese have displayed, of foregoing short term rewards for greater long term successes, should be a key ingredient in this endeavor.

Indulge me in a fantasy, not of the future, but of the past. I should like to think that, if cold fusion had been a burning topic a few years before 1951, as well it might, Nishina would have recognized that it was a subject for open minded research — not suppression. And, in view of the physico-chemical nature of this subject, that he would have thrown all the resources of the institute of which he was president — Riken (Rikagaku Kenkyusho), the Institute of Physical and Chemical Research — into the study and development of cold fusion. Dare one hope that a dream of the past also contains a glimpse of the future?

Domo arigato gozaimasu. Thank you very much.

High Temperature Superconductivity: History and General Review

V.L. Ginzburg

P.N. Lebedev Physical Institute, USSR Academy of Sciences, Leninsky Prospect 53, 117924, Moscow, USSR

Abstract. A great many scientific papers have been devoted to high-temperature superconductors (HTSC). For this reason, in my report I only touch upon several questions. Among these questions are the history of the study of superconductivity and the discovery of HTSC; the calculation of the critical temperature T_c of a superconducting transition and the ways of increasing this temperature; the mechanism providing high T_c values. I also consider the specificities of HTSC in the framework of the macroscopic theory of superconductivity and the circulational thermoconductivity in HTSC. A few remarks concern the future study of the problem.

1. Introduction

High-temperature superconductors (HTSC) are considered to have been discovered in 1986. Although little time has passed since then, a great many original papers and a number of reviews have already been devoted to the problem of HTSC (see, in particular, [1–3]). Proceedings of many, sometimes long and crowded conferences on HTSC have been published (see e.g. [4]). In this situation, I can but briefly elucidate only a small part of the material available. Section 2 includes some remarks of historical character. Section 3 discusses the factors determining the critical temperature T_c of a superconducting transition. Then we consider the mechanisms of superconductivity and the possible nature of high T_c values in HTSC (Sect. 4). Irrespective of the study of the HTSC mechanism, it is necessary to apply the macroscopic theory of superconductivity. The specificity of this theory in application to HTSC is analyzed in Sect. 5. In Sect. 6 I make a remark concerning the thermocirculation effect in HTSC. Finally, in Sect. 7 I touch upon the future and, in particular, the problem of room-temperature superconductors (RTSC). I should note that the present paper overlaps to some extent (except for Sects. 5 and 6) with a more detailed paper [5] written three years ago but published only in 1989. As far as Sects. 5 and 6 are concerned, they are mostly based on my own recent papers. This, I hope, will justify their inclusion in the present report.

Springer Proceedings in Physics, Volume 57
Evolutionary Trends in the Physical Sciences Eds.: M. Suzuki and R. Kubo

2. Remarks of Historical Character

The birth of low-temperature physics can be reasonably associated with helium liquefaction (1908) and with the discovery of superconductivity (1911; both these achievements are due to Kamerlingh Onnes). It is of interest that for 15 years (till 1923) liquid helium had been obtained in Leiden only. Such was the speed of development of science at those times. Today's speed of development is clearly seen by the example of HTSC studies. What will happen in 50–100 years?

For already nearly 80 years the study of superconductivity has been carried out in various directions, namely, physics, obtaining new materials, and technical applications. The history of physical research is elucidated in the first paper of the collection [3]. Here I dwell only on some points, and first of all on the data on the critical temperature T_c of a superconducting transition. The main landmarks are listed in Table 1.

Table 1. Increase of the highest known T_c

Material	T_c $[K]$	The year of the discovery of superconductivity
Hg	4.1	1911
Pb	7.2	1913
Nb	9.2	1930
Nb_3Sn	18.1	1954
Nb_3Ge	23.9	1973

I shall also recall that under atmospheric pressure the boiling temperatures of He, H_2, Ne and N_2 are respectively: $T_{b,He} = 4.2\,K$, $T_{b,H_2} = 20.3\,K$, $T_{b,Ne} = 27.2\,K$, and $T_{b,N_2} = 77.4\,K$.

In the period from 1954 to 1985, when superconductors were already intensively investigated and used, the T_c values increased only by approximately 5 K. Therefore, in spite of the fact that by 1986 there already existed many theoretical arguments in favour of the possibility of creating even "genuine" HTSC with $T_c > T_{b,N_2} = 77.4\,K$ (see [6] and the literature cited there), the HTSC problem had been in the shade. I permit myself, although I have done so already in [5.7], to quote from one of my papers published in 1984 in Russian, giving my view of the situation:

"It somehow happened that research in high-temperature superconductivity became unfashionable (there is good reason to speak of fashion in this context, since fashion sometimes plays a significant part in research work and in the scientific community). It is hard to achieve anything by making admonitions. Typically it is some obvious success (or reports of success, even if erroneous) that can radically and rapidly reverse attitudes. When they sense a 'rich strike' the former doubters, and even dedicated critics, are capable of turning coat and becoming ardent supporters of the new work. But this subject belongs to the psychology and sociology of science and technology, and I shall not dwell on it here. In short, the search for high-temperature superconductivity can readily lead to unexpected results and discoveries, especially since the predictions of the existing theory are rather vague."

I had not, of course, expected this "prediction" to come true so soon. The experience of theoretical HTSC studies for the last 4 years, in spite of the great effort made in this direction, has shown clearly how difficult it is not only to calculate T_c (even approximately) for composite materials, but even to establish the mechanism of superconductivity in these materials. Therefore, theoreticians could hardly supply experimenters with a prediction of how and where HTSC could be better and more reliably sought than was done in [6] (in this connection see also [5]). An exception, I think, is insufficient attention given to superconductivity in $BaPb_{1-x}Bi_xO_3$ (BPBO) discovered in 1974. For this material for $x = 0.25$ we have $T_c \approx 13\,K$, which is too high compared with the T_c value estimated for conventional superconductors. In a related oxide, $Ba_{0.6}K_{0.4}BiO_3$ (BKBO), superconductivity with $T_c \approx 30\,K$ was discovered in 1988. And what is important is that to oxides there also belongs the system $La_{2-x}Ba_xCuO_4$ (LBCO), in which superconductivity with $T_c \sim 30$–$40\,K$ was revealed in 1986 [8], which is taken as the discovery of HTSC. The paper by *Bednorz* and *Müller* [8] has won general recognition (including the 1987 Nobel Prize), so I shall not underestimate its significance (and I am, of course, far from this intention) by mentioning that metal oxides $La_{2-x}(Ba,Sr)_xCuO_4$ had been obtained before in the USSR, France, and Japan. Moreover, at least in the USSR, the conductivity of the oxide $La_{1.8}Sr_{0.2}CuO_4$ had been investigated in 1978 even in liquid nitrogen (for the corresponding reference see [5]). But this was, of course, not the discovery of superconductivity since in this case $T_c \approx 36\,K$. This is a didactic story.

The term "high-temperature superconductivity" was applied earlier even in the case of materials like Nb_3Sn with $T_c \lesssim 20\,K$. Now nearly all metal oxides with $T_c \gtrsim 20\,K$ are called HTSC. I think that it would be most reasonable to apply the term HTSC only to superconductors with $T_c > T_{b,N_2} = 77.4\,K$, discovered for the example of $YBa_2\,Cu_3\,O_{7-\delta}$ (YBCO $\equiv$ 123) at the beginning of 1987. Terminology is, of course, a conditional and inessential thing, but it is just the possibility of working with superconductors cooled by liquid nitrogen that undoubtedly gave rise to great expectations of important technical applications.

At present, a considerable number of HTSC are known. The highest reliably established T_c value, $T_c \approx 125\,K$, has been reached for $Tl_2Ba_2Ca_2Cu_3O_{10}$. For nearly three years now there have appeared dozens of papers reporting the discovery of HTSC with $T_c > 150\,K$ and up to $T_c \sim 300\,K$. But, if not erroneous, these results always appear to be for nonequilibrium and nonreproducible materials. The most recent data known to me (October 1990) refer to the Tl–Sr–V–O system for which $T_c(R = 0) = 132\,K$ [9]. This fact however requires verification in other laboratories. The most significant feature here is the replacement of Cu by V, since all the "genuine" HTSC (i.e. those with $T_c > T_{b,N_2}$) known up to now have contained Cu.

It is noteworthy that a number of papers published beginning in 1978 have mentioned the possibility of the inclusion of HTSC phase in CuCl and CdS. The diamagnetic effects observed were nonreproducible, and it remained unclear whether HTSC was really observed. It seems to me rather probable, however, that HTSC was in fact observed.

Superconductivity is so to say a very delicate phenomenon. This is evident already from the fact that the first relatively complete microtheory of superconductivity, although only for a model, was formulated only in 1957, that is, 46 years after the discovery of superconductivity (I mean the Bardeen-Cooper-Schrieffer or BCS theory). Erroneous is, however, a rather wide-spread opinion that "pairs" with a change $2e$ were first considered by Cooper in 1956. "Pairs" and their Bose-Einstein condensation as the reason for the appearance of superconductivity were first mentioned, to the best of my knowledge, by Ogg as far back as 1946. More important and realistic was the consideration of pairs by Schafroth in 1954 (see reference in [5]). As a matter of fact, Schafroth proposed a model of superconductivity with "local pairs". In this model the pair size ξ_0 is of the order of the atomic scale d, pairs exist also at $T > T_c$, and T_c is the temperature of Bose-Einstein pair condensation. The undoubted success of the BCS model and theory with "large" pairs (i.e. with $\xi_0 \gg d$) so to say eclipsed the Schafroth model. The discovery of superconductivity changed the situation. We shall speak of it below.

3. Critical Temperature T_c in the BCS and Schafroth Models

The BCS model deals with an electron Fermi liquid or, strictly speaking, Fermi gas. The BCS critical temperature is

$$T_c = \Theta \exp(-1/\lambda_{eff}) . \quad (1)$$

Here $k_B\Theta$ is the energy range near the Fermi surface in which there occurs attraction between electrons with opposite spins leading to pair production; λ_{eff} is a dimensionless parameter characterizing attraction in the indicated range. In the BCS model, (1) refers only to the case of weak coupling, when

$$\lambda_{eff} \ll 1 . \quad (2)$$

In the simplest case,

$$\lambda_{eff} = N(0)V , \quad (3)$$

where $N(0)$ is the density of states on the Fermi surface (in the normal state) and V is the matrix element of the interaction energy.

The requirement of the Organizing Committee of the Conference to submit only a short paper deprives me of the possibility of dwelling even briefly on the extension of formula (1) to the case of strong coupling and on generalization of the expression (3). To some extent all this has been done in [5] and in the literature cited there, particularly in [6]. My consideration will be based on formulae (1) and (3) and is presented in a simplified form.

In conventional superconductors (with $T_c < 10$–$20\,\mathrm{K}$), interelectron attraction is assumed to be due to their interaction with phonons (i.e. with the lattice). Then in (1), $\Theta \sim \Theta_D$, where Θ_D is the Debye temperature. Obviously, for $\Theta_D \lesssim 500\,\mathrm{K}$ and $\lambda_{eff} \lesssim 1/3$ the temperature $T_c \lesssim 25\,\mathrm{K}$. This is the usual explanation of why in most cases the T_c value is comparatively low. A more thorough analysis for relatively simple metals confirms this estimate even for strong coupling ($\lambda_{eff} \gtrsim 1$). An exception is metallic hydrogen for which $\Theta_D \sim 3000\,\mathrm{K}$. For compounds of the type of the known "genuine" HTSC the excitation spectrum (in the case of weak coupling these are phonons and electrons or corresponding quasiparticles) is unknown and T_c cannot be calculated within a BCS-type theory. But on the basis of the expressions (1) and (3) we can already outline certain tendencies. Namely, T_c increases with increasing Θ, $N(0)$ and V. As a matter of fact, these parameters are not, of course, independent, but for rough qualitative considerations we shall abstract ourselves from this fact. An increase of $N(0)$ can be reached near structural phase transitions (Yu. Kopaev et al. [5, 6]). This way is promising and is apparently directly related to a number of investigated HTSC. The parameter V increases as the phonon–electron coupling becomes stronger. In the limit of very strong coupling this leads to "local pairs", when the BCS model cannot be applied, and we go over to the Schafroth model. In the simplest case (an ideal Bose gas) in this model

$$T_c = \frac{3.31\hbar^2 n^{2/3}}{m^* k_B} = 2.9 \times 10^{-11} \left(\frac{m}{m^*}\right) \left(n_{\mathrm{cm}^{-3}}\right)^{2/3} \mathrm{K}\,, \tag{4}$$

where m^* is the mass of a pair (a zero-spin boson), m is the free electron mass and n is the pair concentration.

The third possibility, which is clear from (1), is to raise the temperature Θ. In the framework of the phonon mechanism, when $\Theta \sim \Theta_D$, the possibility of increasing Θ is obviously limited. This was the reason for the emergence of the idea (popularized by Little and myself beginning in 1964; see Refs. [5, 6]) to try to "replace" phonons by electron excitons, i.e. by excitations in the subsystem of bound electrons in a metal. This is just the exciton mechanism of superconductivity. For electronic excitons (below called simply excitons) $\Theta \sim \Theta_e \lesssim \Theta_F$, where $E_F = k_B\Theta_F$ is the Fermi energy. In the literature, the exciton mechanism is frequently associated with an account of charge fluctuations. Obviously, the temperature Θ_e can reach, say, $10^4\,\mathrm{K} \approx 1\,\mathrm{eV}$. If λ_{eff} does not decrease strongly, T_c may take on high values. For example, for $\lambda_{eff} = 1/3$ and $\Theta = \Theta_e = 2000$, according to (1) $T_c \approx 100\,\mathrm{K}$. Note that under the weak coupling condition (2) in the BCS theory

$$2\Delta(0)/k_B T_c = 3.53\,, \tag{5}$$

$$\Delta C = 1.43\gamma T_c\,, \tag{6}$$

where $2\Delta(0)$ is the gap in the excitation spectrum in the superconductor at $T = 0$,

ΔC is the jump of the specific heat at T_c and γT_c is the electron part of the specific heat in the normal state at $T = T_c$.

It has been stated that the requirement of metal stability in BCS theory restricts T_c values. But this is not true, and there exist no restrictions of that kind [6]. Thus, HTSC is, in principle, possible in the BCS model even with weak coupling and a sufficiently high, but yet quite realistic Θ value. But the realization of such a possibility is not simple and maybe even impossible in accessible materials.

4. The Mechanism and Character of Superconductivity in HTSC

All the HTSC known at present, in any case Cu-containing ones, possess a complicated structure, are far from being always accessible as good single crystals and are sensitive to the composition. Many of their properties and especially the surface cannot be properly controlled. As a result, the experimental data on HTSC are incomplete and often contradictory. For this reason and owing to the present state of the theory the questions of the nature and, say, the pairing mechanism in HTSC remain disputable and, in effect, open. Competing are the phonon, exciton and magnetic (spin) pairing mechanisms. In the latter case (called also pairing due to spin fluctuations) we mean, figuratively speaking, the exchange by virtual spin waves leading to pair creation. Also competing are the BCS and Schafroth models. The latter case involves in particular consideration of small bipolarons produced in a strong electron–phonon interaction (see [10]). Finally, the so-called RVB model based on the spin liquid concept was proposed. In the normal state this liquid differs radically from a Fermi liquid and, as stated, may become superconducting. I do not understand this mechanism of superconductivity completely (for some data on this model see [2], Chap. 9). Here I think it worth noting that the difference between a spin liquid and Fermi liquid can, in principle, be established in experiment [11, 12]. It is only from comparison with experiment that the role of various mecanisms and the applicability of models can be established.

For oxides $Ba_{1-x}K_xBiO_3$ (BKBO) with $T_c \lesssim 30\,K$ and for the compound $Nd_{2-x}Ce_xCuO_{4-\delta}$ (NCCO) with $T_c \sim 20\,K$ there is every reason to assume the pairing mechanism to be for the most part phonon with the so-called intermediate coupling (the electron–phonon interaction constant $\lambda \sim 1$; $2\Delta(0)/k_B T_c \approx 3.8$–$3.9$; see [13]). A certain role can, of course, be played also by the exciton mechanism. Generally, a more or less rigorous division into phonon and electron subsystems in a metal is possible only in the case of weak coupling ($\lambda \ll 1$). Then in the isotropic BCS model the relations (5) and (6) must hold. For the phonon mechanism, a substantial isotopic effect must, generally speaking, be observed, while in the case of a purely exciton mechanism there is no reason to expect the isotopic effect. A BCS-type excitonic mechanism with $\lambda \ll 1$

underlies the "simplest" HTSC model (see [14] and [7]). True, the comparison with this model is hampered by the fact that the majority of HTSC are layered and strongly anisotropic. So, I shall only remark here that according to some data (see [1], Vol. II, Chap. 2 and [15]) the relation (6) holds well for a number of HTSC. As regards relation (5), it cannot be directly applied to an anisotropic material. The analysis shows that for the phonon–exciton mechanism for HTSC the coupling is either intermediate or strong ($\lambda \gtrsim 1$). Under such conditions we can speak, if you like, of electron–phonon liquid in a metal.

As far as I know, there are no data testifying convincingly against the explanation of the properties of all the known HTSC within the framework of the BCS model with phonon–exciton coupling. But this certainly does not disprove the possibility of another HTSC mechanism not only in general but for the known materials as well. So, for Cu-containing oxides "under suspicion" is the magnetic (spin) mechanism, although it is possible that all the observed magnetic effects are accompanying. In favour of the Schafroth-type models is the fact that pairs in HTSC are comparatively small. But in the Cu-O plane their size is considerably larger then $d \sim 3 \times 10^{-8}$ cm. The true distinction between the Schafroth and BCS models is that in the former pairs exist (not only in the form of fluctuations, but also in a stable form) even above T_c. The indicated data on the specific heat jump at T_c [closeness of the value of the jump to (6)] testify against the Schafroth model (in this case, as in He II, we could expect the jump of specific heat to be much larger than the observed one). The confirmation of the absence of stable pairs at $T > T_c$ would be a convincing disproof of the Schafroth model.

Thus, the establishment of the character of HTSC is to a great extent still waiting. At the same time, what has already been done extends our horizon: clearly, we cannot restrict our consideration to the BCS model with the phonon pairing mechanism for there are many other possibilities.

5. The Macroscopic Theory for HTSC

Important as the microtheory of superconductivity and the HTSC microtheory in particular are, for solving a wide range of problems, especially electrodynamical ones, the macroscopic theory of superconductivity is needed. Such a theory successfully applicable near T_c was developed in 1950 (with extension to the anisotropic case in 1952) [16]. It seems that this theory, in which the complex scalar function Ψ is used as the order parameter, must be fully applicable also in the case of HTSC. This is generally true but with serious stipulations. First, a more complicated quantity than Ψ may appear to be the order parameter (in such cases we speak of "unconventional pairing", see [1], Vol. II, Chap. 9). This is the case with superfluid ^{3}He phases and evidently at least with some heavy fermion superconductors (UPt_3 and others). For HTSC the unconventional pairing is possible but according to all the data available (especially for YBCO = 123, which is at present the most thoroughly investigated HTSC) the pairing is conventional (i.e. s-pairing, where the order parameter is just the complex scalar Ψ).

We henceforth assume that we are dealing with s-pairing. Second, in the known HTSC the coherence length $\xi_0 \equiv \xi(0) \equiv \xi(T=0)$ extrapolated to $T=0$ is very small as distinct from conventional superconductors for which $\xi_0 \gg d \sim 10^{-8}$–10^{-7} cm (for first-order superconductors even $\xi_0 \sim 10^{-4}$ cm). So, according to some data for YBCO, $\xi_{0,z} \equiv \xi_{0,c} \equiv \xi_{0,\perp} \sim 5$ Å and $\xi_{0,ab} \equiv \xi_{0,\parallel} \sim 20$–$30$ Å (here the z- or c-axis is perpendicular to Cu–O layers, and the a- and b-axes lie in the plane of these layers). The Ψ-theory of superconductivity [16] is valid only if $\xi(T) \gg 1$. Close enough to T_c this condition is met [recall that in the Ψ-theory $\xi_l(T) \sim \sqrt{T_c/(T_c - T)}$], but the domain of applicability of the theory becomes narrower. One of the possible generalizations of the theory is connected with the use of the order parameter Ψ only for layers (the two-dimensional case) with allowance made for Josephson interaction between layers. In the domain of applicability of the Ψ-theory, the free energy density has the form [16, 17]

$$F = F_{n0} + \frac{H^2}{8\pi} + a|\Psi|^2 + \frac{b}{2}|\Psi|^4 + \frac{1}{4m_l^*}\left|\left(-i\hbar\nabla_l - \frac{2e}{c}A_l\right)\Psi\right|^2 , \tag{7}$$

where $\boldsymbol{H} = \text{rot}\,\boldsymbol{A}$ is the magnetic field (or, more precisely, magnetic induction) vector, F_{n0} is the equilibrium free energy in the normal state (in the absence of magnetic field), $a = \alpha t$, $b = \text{const}$, $t = (T - T_c)/T_c$, $2m_l^* = \{2m_x^*, 2m_y^*, 2m_z^*\}$ are the principal values of the effective mass tensor of superconducting electron pairs (with a charge $2e$). Obviously, in the isotropic case $m_x^* = m_y^* = m_z^* = m^*$.

On the boundary S of a superconductor with a non-superconductor or with a vacuum the boundary condition has the form

$$n_l \Lambda_l \left(\frac{\partial \Psi}{\partial x_l} - i\frac{2e}{\hbar c} A_l \Psi\right)\Big|_s = -\Psi\Big|_s , \tag{8}$$

where n_l are components of the unit vector $\boldsymbol{n}$ of the normal to the boundary of the specimen and Λ_l are characteristics of the boundary (extrapolation lengths) having the dimension of length. In conventional superconductors, to a good approximation, for the boundary with a dielectric (vacuum) $\Lambda_l(T) \gg \xi_l(T)$, and a much simpler condition is used:

$$\boldsymbol{n}\left(\nabla\Psi - i\frac{2e}{\hbar c}\boldsymbol{A}\Psi\right)\Big|_s = 0 . \tag{9}$$

In the opposite limiting case realized in helium II on the boundary $\Psi|_s = 0$ [18]. The parameter $\Lambda_l \sim \xi_{l,0}^2/d$, and therefore in HTSC, owing to the smallness of $\xi_{l,0} \equiv \xi_l(0)$ the parameters Λ_l are relatively small. This leads, generally speaking, to the necessity of using the boundary condition (8) containing, unlike (9), the quantities $\Lambda_l(T)$, which are not known in advance. Owing to the smallness of $\xi_{0,l}$ the fluctuations in HTSC are large. The point is that the temperature region of strong fluctuations (the critical region) near T_c is proportional to $(\xi_{0,x}^2\, \xi_{0,y}^2\, \xi_{0,z})^{-1}$. As a result, the critical region for HTSC can be substantial even for the three-dimensional case. In the critical region the expression (7) is inapplicable but can be generalized [17] by analogy with He II [18].

Thus, the macroscopic theory of HTSC has an obvious specificity and I am sure will be the subject of numerous studies.

6. On the Thermocircular Effect in HTSC

For a long time thermoelectric effects have been thought of as completely absent in the superconducting state [19]. This is in fact not the case, but the thermoelectric effects in the superconducting state do differ radically from those in the normal state of a conductor. The point is that two currents – superconducting (current density $\boldsymbol{j}_s$) and normal (current density $\boldsymbol{j}_n$) – can flow in a superconducting state. The situation is similar to that in the superfluid liquid (in particular, in He II), where superfluid and normal liquid flows can exist. We shall not dwell here in general on thermoelectric effects in the superconducting state, which were noticed as far back as 1944 [20] but have not yet been thoroughly investigated. We shall discuss only one phenomenon, namely, the thermocircular heat conductivity in the superconducting state. Let us consider a nonuniformly heated superconducting rod (the temperature gradient along the rod is ∇T). In the normal state, the current density $\boldsymbol{j}$ in the rod is certainly equal to zero (the rod is not closed), and on its ends there occurs a certain thermoelectromotive force ε. In the superconducting state $\boldsymbol{j} = 0$, but in the rod the currents with densities $\boldsymbol{j}_{\mathrm{n}} = b_{\mathrm{n}}\nabla T = -\boldsymbol{j}_{\mathrm{s}}$ flow towards each other. Of course, the total current $\boldsymbol{j} = \boldsymbol{j}_{\mathrm{s}} + \boldsymbol{j}_{\mathrm{n}}$ and the magnetic field $\boldsymbol{H} = 0$ (the rod is assumed to be isotropic and uniform; for more details see [21]). At the ends of the rod the currents $\boldsymbol{j}_{\mathrm{s}}$ and $\boldsymbol{j}_{\mathrm{n}}$ transform into one another, which causes an additional (thermocirculatory) heat transfer due to pair creation or destruction. Therefore, in a superconducting state the coefficient of thermoconductivity $\kappa = \kappa_{\mathrm{ph}} + \kappa_{\mathrm{el}} + \kappa_{\mathrm{c}}$, where κ_{ph}, κ_{el} and κ_{c} correspond respectively to contributions to the heat conductivity due to the lattice (phonons), normal electrons and circulation.

The estimation gives

$$\kappa_{\mathrm{c}}/\kappa_{\mathrm{el}} \sim k_{\mathrm{B}}T/E_{\mathrm{F}}\,, \tag{10}$$

where E_{F} is Fermi energy in the metal.

In conventional superconductors $T_{\mathrm{c}} < 10\,\mathrm{K}$ and $E_{\mathrm{F}} \sim 3$–$10\,\mathrm{eV}$ and, therefore, $\kappa_{\mathrm{c}}/\kappa_{\mathrm{el}} < 3\times10^{-4}$, that is, the circulational contribution is negligibly small. But in HTSC, for example, for $T_{\mathrm{c}} \sim 100\,\mathrm{K}$ and $E_{\mathrm{F}} \sim 0.1$–$0.3\,\mathrm{eV}$ we have already $\kappa_{\mathrm{c}}/\kappa_{\mathrm{el}} \sim 0.03$–$0.1$. The ratio $\kappa_{\mathrm{c}}/\kappa_{\mathrm{el}}$ can, in fact, be much larger [the estimate (10) is rather rough], and upon unconventional pairing $\kappa_{\mathrm{c}}/\kappa_{\mathrm{el}} \sim 1$ even, according to a rough estimate. Hence, in HTSC (as well as in superconductors with heavy fermions) the thermocircular effect can be considerable. The interesting question of thermoelectric phenomena in superconductors is mentioned here, I should confess, also for a subjective reason – both the last and one of my first papers on superconductivity (written with an interval of 45 years) were devoted to it [20, 21].

7. On the Future (HTSC and RTSC)

Before 1987 not much attention had been paid to the problem of HTSC, but for the last four years the situation has changed radically. We could, perhaps, even be surprised that in spite of extensive efforts the progress in physics and applications of HTSC is rather modest. It is no wonder, however, in view of some specificities of the known HTSC. At the same time, the progress in the experimental technique of HTSC studies is impressive (see e.g. [12, 13]). I think that in another four years those HTSC which are known now will be fairly well investigated and the main unclear physical questions (the model of a superconductor, the character and mechanism of pairing, etc.) solved. Without doubt, the fields of application of HTSC in technology, medicine, etc. will also gradually widen.

Among the problems, I would single out the maximal possible critical temperature value $T_{c,max}$ (I mean not very exotic substances, of the type of metallic hydrogen, which are stable under atmospheric pressure). In [6] we gave the estimate $T_{c,max} \lesssim 300\,K$ because of the absence of any known essential restrictions on T_c. Since then nothing has changed in the theory, but the experimental progress (obtaining HTSC with $T_c \lesssim 130\,K$) suggests the estimate

$$T_{c,max} \sim 400 - 500\,K\,. \tag{11}$$

Thus, we may speak already not of HTSC but of RTSC (room-temperature superconductors.

The estimate (11) is intuitive and has been made without any deep reason. If RTSC are created, they will be "cooled" with water, which would be a jump comparable with going over from cooling with liquid He to that with liquid N_2. HTSC with $T_c \sim 200$–$250\,K$ can also be of some technological interest since in this case liquid N_2 can be replaced by other coolants.

Is it realistic to create RTSC? It is, of course, difficult to answer this question with confidence if we are not guided by a disputable thesis: everything that is not forbidden is allowed. In any case, the RTSC problem has now taken the place that had belonged to HTSC before 1987.

References

1 D.M. Ginsberg (ed.): *Physical Properties of High Temperature Superconductors*, Vols. I, II (World Scientific, Singapore 1989, 1990)

2 J.W. Lynn (ed.): *High-Temperature Superconductivity*, Graduate Texts in Contemporary Physics (Springer, Berlin, Heidelberg 1990)

3 J.G. Bednorz, K.A. Müller (eds.): *Earlier and Recent Aspects of Superconductivity*, Springer Ser. Solid-State Sci., Vol. 90 (Springer, Berlin, Heidelberg 1990)

4 Proc. Int. Conf. Materials and Mechanisms of Superconductivity II. Physica C **162–164** (1989)

5 V.L. Ginzburg: Prog. Low-Temp. Phys. **12**, 3 (1989)

6 V.L. Ginzburg, D.A. Kirzhnits (eds.): *High Temperature Superconductivity* (Consultants Bureau, New York 1982)

7 V.L. Ginzburg: Phys. Today No. 3 (March), 9 (1989)

8 J.G. Bednorz, K.A. Müller: Z. Phys. B **64**, 189 (1986)
9 High T_c Update **4**, No. 19 (1990)
10 A.S. Alexandrov, D.K. Ray: Nature (in press); see also Rev. Mod. Phys. **62**, 113 (1990)
11 P.W. Anderson: Phys. Rev. B **42**, 2624 (1990)
12 C.G. Olsen, R. Liu, D.W. Lynch, R.S. List, A.J. Arko, B.W. Veal, Y.C. Chang, P.Z. Jiang, A.P. Paulikas: Phys. Rev. B **42**, 381 (1990)
13 Q. Huang, J.F. Zasadzinski, N. Tralshawala, K.E. Gray, D.G. Hinks, J.L. Peng, R.L. Greene: Nature **347**, 369 (1990)
14 W. Little: Science **242**, 1390 (1988)
15 J.E. Gordon, M.L. Tan, R.A. Fisher, N.E. Phillips: Solid State Commun. **69**, 625 (1989)
16 V.L. Ginzburg, L.D. Landau: Zh. Exp. Teor. Phys. **20**, 1064 (1950); V.L. Ginzburg: Sov. Phys.–JETP **23**, 236 (1952)
17 V.L. Ginzburg: Physica C **153–155**, 1617 (1988) L.N. Bulaevskii, V.L. Ginzburg, A.A. Sobyanin: Sov. Phys.–JETP **68**, 1499 (1988)
18 V.L. Ginzburg, A.A. Sobyanin: J. Appl. Phys. Jpn. **26**, Suppl. 26–3, 1785 (1987)
19 D. Shoenberg: *Superconductivity* (Cambridge University Press, Cambridge 1965)
20 V.L. Ginzburg: J. Phys. USSR **8**, 148 (1944)
21 V.L. Ginzburg: J. Supercond. **2**, 323 (1989); Supercond. Sci. Technol. (in press, 1990)

SO(4) Symmetry in a Hubbard Model

C.N. Yang

Institute for Theoretical Physics, State University of New York, Stony Brook, NY 11794-3840, USA

Abstract. For a simple Hubbard model on a periodic lattice, we construct operators **J** and **J**' that form two sets of mutually commuting angular momenta operators. Furthermore, all components of **J** and **J**' commute with the Hamiltonian. Thus we have four quantum numbers j, j', j_z, and j_z' for the system. Two of these, j' and j_z', are related to the ODLRO for the model, hence to its possible superconductivity.

1. Introduction

Consider the Hubbard model on an LxLxL periodic lattice, where L=even. The lattice sites will be denoted by $\mathbf{r}$ = (i,j,k), where i,j,k are integers. The momenta will be $\mathbf{k} = (2\pi/L)(I,J,K)$ (mod 2π) where I,J,K are integers. The system clearly conserves momenta. We shall write the annihilation operators for spin up and spin down electrons as $a_\mathbf{r}$ and $b_\mathbf{r}$ and take

$$[a_\mathbf{r}, b_\mathbf{r}]_- = [a_\mathbf{r}, b_\mathbf{r}^+]_- = 0. \tag{1}$$

The operators $a_\mathbf{r}$ and $a_\mathbf{r}^+$ anticommute with each other as usual, so do $b_\mathbf{r}$ and $b_\mathbf{r}^+$. The commutation relation (1) can be chosen because the system conserves the total number of a and b particles. We take the Hamiltonian to be H=T+V where

$$T = \Sigma\text{-}2(\cos k_\mathbf{x} + \cos k_\mathbf{y} + \cos k_\mathbf{z})\,(a_\mathbf{k}^+ a_\mathbf{k} + b_\mathbf{k}^+ b_\mathbf{k}) \tag{2}$$

$$V = 2W\Sigma a_\mathbf{r}^+ a_\mathbf{r} b_\mathbf{r}^+ b_\mathbf{r} \tag{3}$$

and $a_\mathbf{r}, b_\mathbf{r}$ are Fourier transforms of $a_\mathbf{r}$ and $b_\mathbf{r}$:

$$a_\mathbf{k} = L^{-3/2}\,\Sigma a_\mathbf{r} e^{-i\mathbf{k}.\mathbf{r}}\,.$$

Clearly the total number of a and b particles, N_a and N_b, are constants of motion.

First consider the very simple case of $N_a = N_b = 1$, and total momentum $= \boldsymbol{\pi} = (\pi,\pi,\pi)$. The kinetic energy T in such a case is identically equal to zero, because

$$\cos k + \cos(\pi - k) = 0. \tag{4}$$

Springer Proceedings in Physics, Volume 57

Evolutionary Trends in the Physical Sciences Eds.: M. Suzuki and R. Kubo

We thus find the state $\eta^+|0\rangle$ to be an eigenstate of H where

$$\eta = \Sigma a_{\mathbf{k}} b_{\pi-\mathbf{k}} \tag{5}$$

and $|0\rangle$ is the vacuum. Notice that we can also write

$$\eta = \sum_b a_{\mathbf{r}} b_{\mathbf{r}} - \sum_W a_{\mathbf{r}} b_{\mathbf{r}}$$

where we have divided the lattice sites checker-board fashion into black and white sites, and the first sum is over the black sites, the second over the white sites.

The operator [1] η turns out to be very useful for the Hubbard problem. By studying its commutations relations with T and V we find e.g. that

$$(\eta^+)^m |0\rangle$$

is also an eigenstate of H if m is not too large.

η^+ is an operator that pairs a spin up electron with a spin down electron so that their total momentum is π. *This is different from the usual Cooper pairing*, and we shall exploit its usefulness in the rest of this paper.

2. Two Sets of Operators J and J' and SO4 Symmetry

It is more convenient to use a Hamiltonian H' which is *trivially* different from H:

$$H' = T+V',$$

$$V' = 2W\Sigma(a_{\mathbf{r}}^+a_{\mathbf{r}} - (1/2))(b_{\mathbf{r}}^+b_{\mathbf{r}} - (1/2)) = V-W(N_a+N_b)+(1/2)WL^3. \tag{6}$$

We define J_x' and J_y' by [2]

$$\eta = J_x'-iJ_y', \quad \eta^+ = J_x'+iJ_y' \quad \text{and}$$

$$J_z' = -i[J_x',J_y'] = (1/2)(N_a+N_b)-(1/2)M, \; M = L^3. \tag{7}$$

Then J_x',J_y' and J_z' commute with each other like the components of angular momenta, and each commutes with T and V'. [Notice we have switched the notation for J and J' here from that used in reference 2.]

Furthermore [2] if we define

$$\mathcal{Y} = \Sigma a_{\mathbf{r}} b_{\mathbf{r}}^+, \quad \text{and}$$

$$\mathcal{Y} = J_x-iJ_y, \mathcal{Y}^+ = J_x+iJ_y,$$

$$J_z = -i[J_x,J_y] = (1/2)(N_a-N_b) \tag{8}$$

then J_x,J_y and J_z commute with each other like the components of angular momenta, and each commutes with T and V'. Furthermore every component of **J** commutes with every component of J'.

Thus we have quantum numbers j and j' where

$$\mathbf{J}^2 = j(j+1), \mathbf{J}'^2=j'(j'+1),$$

and also quantum numbers j_z and j_z'. The eigenstates of H' form into multiplets (j,j'), with all states in the same multiplet having the same eigenvalue for H'.

It is obvious from (7) and (8) that $j_z'+j_z = N_a - (1/2)M$ = integer. Thus

$$j+j' = \text{integer}.$$

It follows from this that [3] the system has SO_4 symmetry.

3. ODLRO

In 1962 it was pointed out [4] that the concept of ODLRO is important for superconductivity. Specifically

ODLRO → flux quantization in units ch/2e
→ Meissner effect
→ superconductivity.

It can be shown that the quantum numbers j' and j'_z are related to the existence of ODLRO. In particular, if $j'^2-j_z'^2$ is of the order of $M = L^3$, the system exhibits ODLRO. I refer the reader to reference 2 for a proof of this statement.

4. Remarks

The considerations above are valid for any dimension. It is only important that the lattice sites should allow a division into black and white sites, with a properly chosen Hamiltonian operator.

The question whether the Hubbard model under consideration exhibits superconductivity at low temperatures for some values of the coupling constant W reduces to the question of the quantum numbers j' and j'_z for low lying states. I am pursuing this approach for the case of 2 and 3 dimensions.

For 1 dimension, the Hubbard model defined by (2) and (3) has been solved by E. H. Lieb and F. Y. Wu [5], using the method of C. N. Yang [6]. With the introduction of the operators **J** and **J'** a natural question arises: what are the quantum numbers j and j'_z for Lieb and Wu's solutions? This is a very good, but not so simple mathematical problem.

Generalization of the operators **J** and **J**' to more complicated Hubbard models will be presented elsewhere.

Hamilton (6) does not *transparently* possess SO4 symmetry. Is it possible to render this symmetry transparent? The answer is yes. By suitably changing the dynamical variables from a, a^+, b and b^+ to a new set of anticommuting invariant. Details will be published elsewhere.

The reader is referred to an interesting paper by S. C. Zhang [7] which discussed possible experimental tests for SO4 symmetry.

References

1. Chen Ning Yang, Phys. Rev. Letters 63, 2144 (1989).
2. Chen Ning Yang and S. C. Zhang, Mod. Phys. Letters B4, 759 (1990).
3. See reference 2 above, and also M. Pernici, Europhysics Letters 12 (1), 75 (1990).
4. C. N. Yang, Rev. Mod. Phys. 34, 694 (1962).
5. E. H. Lieb and F. Y. Wu, Phys. Rev. Lett. 20, 1445 (1968).
6. C. N. Yang, Phys. Rev. Lett. 19, 1312 (1967).
7. Shoucheng Zhang, Phys. Rev. Lett. 65, 120 (1990).

Some Trends in Solid State Physics

Yu.A. Ossipyan

Institute of Solid State Physics, USSR Academy of Sciences,
142432, Chernogolovka, Moscow District, USSR

A most active period of Professor Nishina's life in science fell within the time period when the principles of quantum theory began to be confirmed in experiments with microparticles. Much was said about it in the first two days of our symposium. In my report, devoted to certain aspects of solid state physics, I should like to dwell upon only two new fields developed in recent years. Both these fields involve a clear manifestation of quantum laws. To be more exact, in the two phenomena about which I shall be speaking the role of quantum laws is not that there appear quantitative corrections to the values of quantities entering into particular equations, but that new physical phenomena themselves appear and turn out to be observable solely due to the existence of quantum laws.

I shall be concerned with the following phenomena:

1. Quantum Hall effect
2. Quantum diffusion and crystallization.

So, the term "quantum" is present in the name of both phenomena.

The first of these, the quantum Hall effect (QHE), is related to manifestation of quantization laws in the behaviour of a 2D electron system existing inside a semiconductor crystal. Then there come quantum diffusion and quantum crystallization, related to the appearance in helium crystals that are in equilibrium with superfluid liquid helium of specific defects of purely quantum nature. We are dealing here with manifestation of quantum laws controlling the behaviour of the system of comparatively light helium atoms rather than that of the electron system.

1. Quantum Hall Effect

One of the brightest events in semiconductor physics was the discovery of Hall resistance quantization, termed the quantum Hall effect (QHE) [1]. For this discovery K. von Klitzing was awarded the Nobel prize in physics in 1985. This effect is observable in two-dimensional (2D) electron and hole layers of the space charge in MOSFET structures and modulation-doped heterojunctions. The essence of the experimentally observable phenomenon consists in that, in a low temperature region ($T \sim 1$ K) in the presence of a strong magnetic field, the dependence of the Hall conductivity σ_H on the field H or on the concentration of 2D-carriers

Springer Proceedings in Physics, Volume 57
Evolutionary Trends in the Physical Sciences Eds.: M. Suzuki and R. Kubo

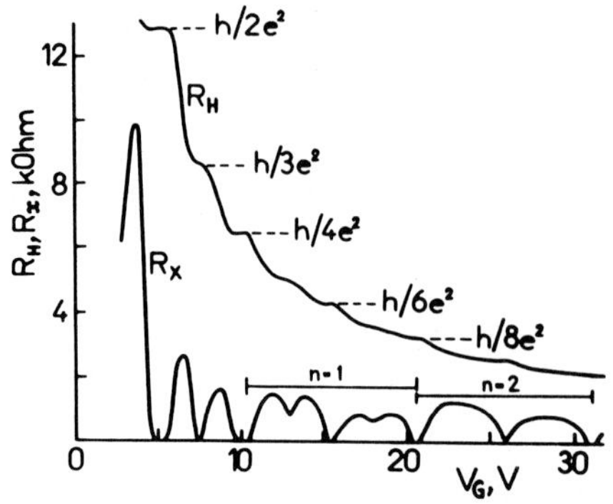

Fig. 1. Quantum Hall effect in a Si-MOSFET structure. Hall resistance R_{H} and resistivity R_x vs the gate voltage V_{G} in a rectangular MOSFET structure. The voltage V_{G} is proportional to N_{s}. The field $H = 18.9\,\mathrm{T}$, $T = 1.5\,\mathrm{K}$

N_{s} exhibits a number of plateaux. This is shown in Fig. 1. Here $R_{\mathrm{H}} \approx \sigma_{\mathrm{H}}^{-1}$ is the Hall resistance. On the plateau

$$\sigma_{\mathrm{H}} = \nu \frac{e^2}{\hbar}, \tag{1}$$

where e is the elementary charge and $\hbar$ is the Planck constant, i.e. fundamental constants, ν is the numerical coefficient termed the filling factor

$$\nu = \frac{N_{\mathrm{s}}}{N_{\mathrm{L}}}, \tag{2}$$

N_{L} is the number of states on the Landau level per unit surface, and, if λ is the magnetic length $\lambda = (e\hbar/eH)^{1/2}$, and $\phi_0 = ch/e$ is the magnetic flux quantum, then

$$N_{\mathrm{L}} = \frac{1}{2\pi\lambda^2} = \frac{eH}{ch} = \frac{H}{\phi_0}. \tag{3}$$

An outstanding feature of the QHE measurement results was that the ν value, corresponding to different plateaux, turned out exactly integral (Fig. 1). It should be noted that the plateau values for σ_{H} correlate with deep gaps in the values of the diagonal components of the conductivity tensor (σ_{xx}) or resistivity tensor (ϱ_{xx}), Fig. 1.

It is not wrong to say that the character of these results of today's physics is externally very close to the basic results which appeared and were interpreted in the times of Professor Nishina when the foundations of quantum physics were laid.

The equation (1) for an ideal electron gas can easily be obtained from the following simple considerations. The density of the Hall current $J_{\mathrm{H}} = eN_{\mathrm{s}}V_{\mathrm{H}}$, where V_{H} is the Hall velocity, $V_{\mathrm{H}} = cE/H$, E being the electric field. Taking into account that $J_{\mathrm{H}} = \sigma_{\mathrm{H}}E$, we obtain the usual relationship

$$\sigma_H = ecN_s/H \tag{4}$$

and then (1). However, this formal result does not explain QHE, it, rather, places emphasis on the dificulties associated with its explanation. In fact, according to (4) with $H = \text{const}$, $\sigma_H \sim N_s$, and with $N_s = \text{const}$, $\sigma_H \sim 1/H$, i.e. in either case there are no hints of the presence of plateaux.

As a possible physical mechanism one might propose that plateaux arise as a consequence of trapping of the majority of carriers into localized states, for example, on defects, as a result of which an increase of N_s does not lead to growth of the concentration of electrons conducting the Hall current. But for captured carriers $V_H = 0$, so the current is transferred by free carriers alone, which disagrees with numerical experimental results, reflected in the equations (1) and (2). So, the general consideration of the experimental results suggests only that there arises a peculiar "conductivity quantum", e^2/h. The discovery of QHE has actuated great scientific forces. It involved a great number of physicists and engineers: experimentalists, working in semiconductor physics, metrologists, and theorists – specialists in the theory of solid state and field theory. The latter is especially symptomatic since it shows that interpretation of QHE belongs to fundamental problems.

Now some words about what experimentalists succeeded in observing when they began to advance into the region of large magnetic field values and, specifically, into the lower temperature region. It turned out that along with the integral quantum Hall effect, the picture of Hall resistance quantization may be observed for fractional filling numbers ν (1/3, 2/3, etc.). *Tsui* et al. [2] who worked with perfect GaAs-based heterostructures with high carrier mobility ($\mu \approx 10^5\,\text{cm}^2/\text{Vs}$), were the first to observe it. Figure 2, where these results are presented, demonstrates plateaux in $\varrho_{xy}(H)$ and minima in $\varrho_{xx}(H)$ that precisely correspond to $\nu = 1/3$ and $\nu = 2/3$.

For the past several years the transport properties of 2D electron systems in the region of large magnetic fields and low temperatures have been the subject of intensive studies. Despite the fact that much has been achieved in understanding the QHE nature, many important questions are still unclear.

Leaving aside concrete experimental and theoretical results, in most general form the results of investigations in the field of quantum transport may be summed up as follows [3]:

1. The QHE is a specific 2D effect related to the presence of gaps in an electron spectrum in a magnetic field.
2. The IQHE is by nature a one-electron effect. Its appearance is connected with a special structure of the 2D electron spectrum in a random field: almost all the states are localized, the extended states transfer the current only in the Hall direction.
3. The FQHE, on the other hand, results from the interelectronic interaction. This interaction gives rise to electronic phases which are denser than the ideal electronic gas (electronic liquids). These phases may have additional gaps in the spectrum.
4. Principle limitations to the quantization accuracy are unknown so far.

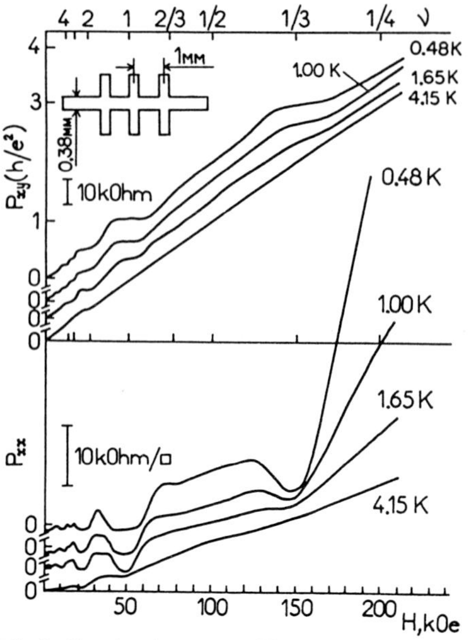

Fig. 2. Fractional quantum Hall effect in a GaAs heterostructure. The sample had $N_s = 1.23 \times 10^{11}\ \mathrm{cm}^{-2}$ and $\mu_e = 9 \times 10^4\ \mathrm{cm}^2/\mathrm{Vs}$

In the course of studies of the phenomena related to the FQHE one would naturally desire to increase the magnetic field and to decrease the temperature in an attempt to bring up the electronic liquid to such a state that electronic crystallization is possible. However, in this case, owing to an increase of the localization degree, it turns out to be practically impossible to investigate the system by methods connected with measurements of transport properties. Great progress has been achieved by using Raman spectroscopy methods for studying 2D electron systems. These methods as applied to 2D systems were developed by *V.B. Timofeev* and his co-workers [4]. Using samples of GaAs-AlGaAs heterostructures they were able to observe jumps of the spectral position of the luminescence line corresponding to recombination of 2D electrons with photoexcited holes. By increasing the magnetic field up to 28 T they recorded the jumps indicating gaps in the energy spectrum of 2D electrons and corresponding to the filling numbers 1/5, 1/7, and even 1/9 (Fig. 3). By replotting the obtained results in jump temperature–filling factor coordinates the authors of these works obtained a specific diagram of state of the 2D system (Fig. 4). They suggested that the gap parameters on the diagram correspond to the crystallization parameters in the 2D electron system in the ultraquantum limit [5]. As for crystallization of 2D electrons into a Wigner crystal in the classical regime, this crystallization was earlier observed in experiments with electrons on the liquid helium surface [6].

In going on to discuss quantum crystallization, I would like to switch over from the electronic system to the system of helium atoms, where quantum properties are manifested rather vividly at low temperatures.

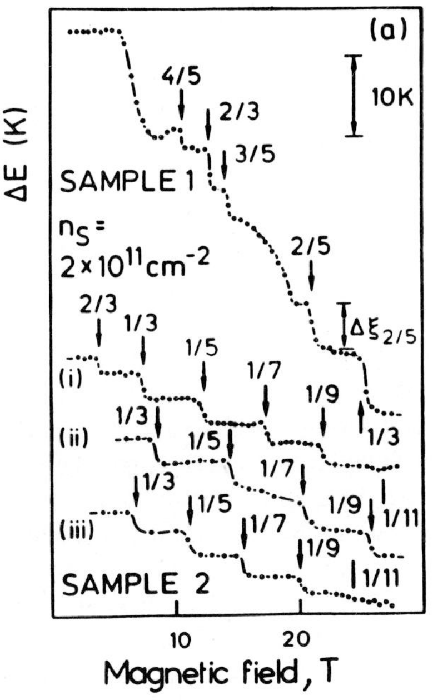

Fig. 3. Jumps of the spectral position of the luminescence line, corresponding to the electron recombination from the lower spin state with photo-excited holes in the δ-doped monolayer, measured with varying magnetic field in samples 1 and 2 for various concentrations: (i) 0.59×10^{11} cm^{-2}; (ii) 0.7×10^{11} cm^{-2}; (iii) 0.54×10^{11} cm^{-2}

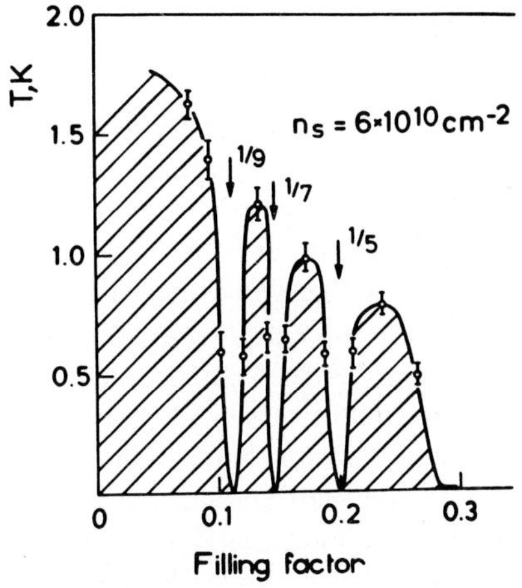

Fig. 4. Phase diagram of the liquid–Wigner crystal transition

2. Quantum Crystals

The most pronounced quantum crystals are crystals of the helium isotopes ^4He and ^3He. At low temperatures the idea of quantum-mechanically indistinguishable identical particles plays a determining role in the properties of liquid helium, since liquid helium is a quantum liquid. Its crystallization occurs at an insignificant increase of pressure (25–30 atm) and is followed by a relatively small (less than 10%) change of the density. The herewith arising crystals retain, to a marked degree, their quantum nature [7].

Among quantum crystals are crystals of hydrogen and its isotopes. The list of quantum crystals containing particles of only one kind is here exhausted. As for crystals containing particles of different kinds, then besides the solutions of helium and hydrogen isotopes there are other interesting crystals which are partly quantum. In these the quantum effect of particle delocalization is of importance not for all but only for some of the particles forming the crystal. This is true, e.g., for hydrogen solutions in lattices of some heavy metals (niobium, zirconium, palladium). Because of small mass and comparatively weak interaction with atoms of the metal matrix, hydrogen atoms are delocalized in the crystal whereas the matrix atoms themselves behave in quite a classical manner.

3. Quantum Diffusion

Elementary consideration of the possibility of quantum diffusion has to allow for the fact that in the simplest case of the crystal consisting of identical atoms the tunneling processes are not directly observable, since they are reduced to exchange of one atom for another exactly the same. Therefore the simplest way to reveal the particle delocalization is as follows:

We shall consider a crystal containing an impurity atom, for example, a ^{4}He crystal doped with ^{3}He. Due to tunneling processes the impurity atom, by exchanging places with environmental matrix atoms, gets delocalized in the crystal. According to quantum mechanics the herewith arising situation turns out to be quite analogous to the well-known situation with electrons in metals. Like electrons, the impurity atoms are transformed into quasiparticles, freely moving through the crystal at a constant velocity. If the impurity concentration is small enough, they behave as a rarefied quasiparticle gas. So, we arrive at the important conclusion that quantum crystals have to demonstrate occurrence of quantum diffusion whose characteristic features are the same as in the gas of freely moving particles. The dependence of the diffusion coefficient on the temperature and concentration of impurities should be quite unusual. In normal crystals the diffusion coefficient decreases monotonically with decreasing temperature, and at not too high concentrations it is independent of these. In quantum crystals the temperature dependence of the diffusion must have the complicated behaviour shown schematically by the solid curve in Fig. 5.

There exist three characteristic temperature regions. In region I at sufficiently low temperatures the free motion of impurity quasiparticles is disturbed only by their collisions with each other. The diffusion coefficient is determined by the

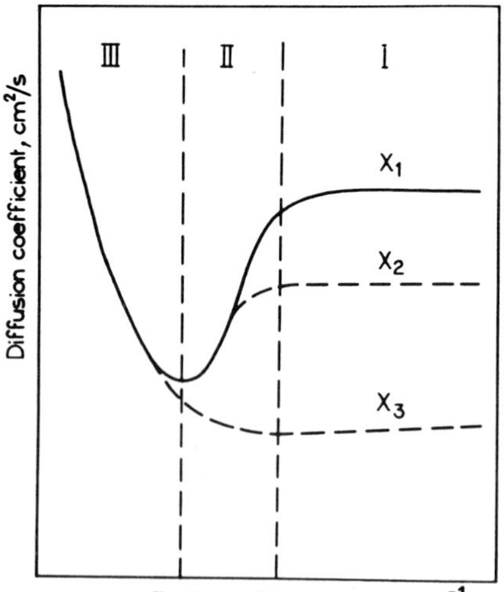

Fig. 5. Temperature dependence of the diffusion coefficient at various impurity concentrations

concentration but is temperature independent. The lower the concentration, the greater the free path of the quasiparticles. Inasmuch as the free motion in the absence of collisions is characterized by an infinite diffusion coefficient, the diffusion coefficient, observable in region I, must grow unboundedly with decreasing concentration. In the region of higher temperatures (II) collisions of impurity quasiparticles with lattice oscillation quanta, i.e. phonons, become significant. As the temperature is increased the number of phonons grows, which leads to a reduction of the free path length and the diffusion coefficient. So, in region II one must observe a decrease of the quantum diffusion with increasing temperature. It is precisely due to this that at sufficiently high temperatures the quantum diffusion should be replaced by the usual diffusion mechanism that, in region III, leads to a growth of diffusion with increasing temperature. Dashed curves in Fig. 5 demonstrate a change of the temperature dependence behaviour with increasing impurity concentration x for $x_3 > x_2 > x_1$.

The experimental data on the temperature dependence of diffusion for various concentrations are demonstrated in Fig. 6. It is seen that the experimental curves are fully coincident in behaviour with the curves of Fig. 5, corresponding to theoretical presentations. A more detailed study shows that there also exists a quantitative agreement between theory and experiment.

Figure 7 demonstrates the data on the concentration dependence of the diffusion of the low temperature region (I). The experimental points fall well on the solid straight line, corresponding to the theoretically predicted dependence. As the concentration decreases the diffusion coefficient grows, which agrees fairly well with the fact of free motion of an isolated impurity quasiparticle.

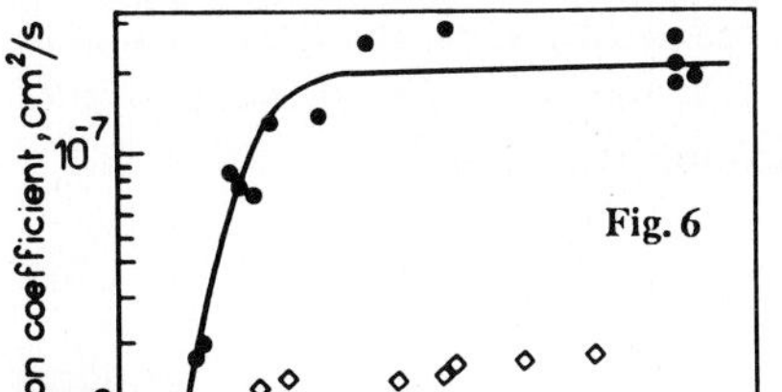

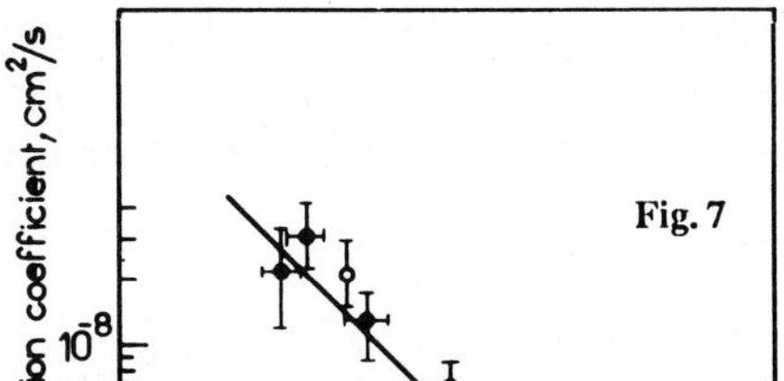

Fig. 6. Temperature dependence of the diffusion coefficient of ^3He in ^4He at various concentrations: •: 6×10^{-3}%, ◇: 5×10^{-2}%, o: 0.75%

Fig. 7. Diffusion coefficient of ^3He in ^4He crystals vs the concentration at lower temperatures

The above experimental data enable a direct determination of such characteristics of the energy spectrum of impurity quasiparticles as the width of their energy band and characteristic velocity. It appears that for ^{3}He crystals these values are equal to 10^{-4} K and 0.1 cm/s, respectively; that is, they are rather small. They are much smaller than all the other characteristic energies and velocities in the crystal. This fact stipulates rather unusual properties of the impurity quasiparticles which, to a considerable extent, retain their sense even in conditions when their free path is much smaller than the lattice period.

Interestingly, recently it has been found that quantum diffusion can be observed in normal crystals as well, but for unusual impurities. I mean the development of a method for measuring the diffusion coefficient of positive muons in crystals. Inasmuch as muons are much lighter than the lightest atoms, quantum tunneling processes are essential for them even in lattices of normal crystals.

The work on quantum diffusion convincingly proves that impurities in quantum crystals behave as delocalized quasiparticles freely moving across the crystal. A very clear manifestation of this is the rapid diffusion growth for a temperature decrease in the phonon region II (see Fig. 6). So, the fact of quantum particle delocalization basic to quantum crystal physics has been proved.

In view of this it should be noted that the full number of lattice sites of a quantum crystal may not coincide with the number of atoms, but exceed it. As a result there forms a quantum crystal in which all the sites are identical but not fully occupied. This implies that the crystal contains zero-point vacancies since these vacancies, like zero-point oscillations of crystals, do not disappear at absolute zero temperature.

Quantum crystals with zero-point vacancies are surprising objects. In these crystals two types of motion are possible simultaneously: one, characteristic for solids, is related to the motion of atoms through lattice sites, the other is due to the motion of zero-point vacancies and possessess the properties of the motion of a liquid. At low temperatures these crystals exhibit superfluidity.

4. Crystallization Waves

The motion of zero-point vacancies allows generalization for other types of defects in crystals. If at sufficiently low temperatures normal crystals always have a characteristic faceting in the equilibrium state then on the surface of quantum crystals regions must necessarily exist where, even at absolute zero temperature, the faceting is violated by the quantum delocalization of steps and kinks. Besides, faceted regions may also exist. Most interesting properties should be demonstrated by unfaceted surfaces of ^{4}He-type crystals, whose boundary at low temperatures represents a boundary with a superfluid component. In this case the motion of the system of zero-point surface defects is in its properties analogous to that of a superfluid liquid. The motion of steps and kinks along the surface is followed by growth or melting of the crystal. Therefore the presence of zero-point defects must lead to an appreciable change in the character of the crystallization and melting processes. These processes, in this case, may be regarded as supercrystallization and supermelting, since their rate is extremely great and they occur coherently, without a significant energy dissipation, i.e. like other quantum superphenomena.

A direct consequence of this presentation is the possibility of propagation of weakly attenuating melting and crystallization waves along the surface. In the external manifestations these waves are similar to capillary waves at the normal liquid–gas interface: both are accompanied by the wavy motion of the interface. For capillary waves, the boundary motion proceeds together with the material, in the absence of the evaporation and condensation processes. However, crystallization waves are totally due to the transition of the material from the liquid phase to the crystalline and back. The crystal itself is at rest, and surface oscillations are related to the fact that in some of its sites the crystal gets crystallized, in others, melted. A simple visual observation of the ^4He crystal surface at low temperatures shows that externally it looks rather like an interface between two low-viscosity immiscible liquids. Under the action of even insignificant vibrations of the experimental setup the surface continuously oscillates.

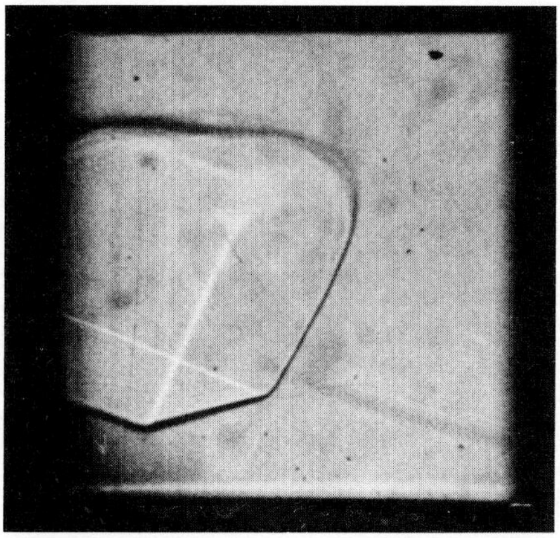

Fig. 8. Crystallization wave on the surface of a partly faceted ^4He crystal

Figure 8 presents a photograph of a ^4He crystal at a temperature of 0.5 K that clearly demonstrates the duality of the quantum crystal properties. In the lower crystal part one can see the faceting, indicating that the sample is a perfect single crystal. On the upper rounded surface one can see a crystallization wave caused by the vibration of the setup. This part of the surface looks like a liquid surface.

Quantum diffusion and the phenomena occurring on the crystal surface, especially crystallization waves, are, at present, the clearest manifestations of the peculiar nature of quantum crystals. The gigantic rate of the processes of crystallization and melting of quantum crystals opens up unique possibilities to crystallography. It enables investigations of some interesting properties common to all crystals but practically inaccessible for studies in normal crystals. These involve all capillary phenomena in solids, particularly the insufficiently studied field of the physics of phase transitions, followed by the appearance or disappearance of crystal faceting.

In conclusion, I should like to express cordial gratitude to Prof. A.F. Andreev and Prof. V.B. Timofeev for very fruitful discussions.

References

1 K. von Klitzing, G. Dorda, M. Pepper: Phys. Rev. Lett. **45**, 494 (1980)
2 D.C. Tsui, H.L. Störmer, A.C. Gossard: Phys. Rev. Lett. **48**, 1559 (1982); R.B. Laughlin: Phys. Rev. Lett. **50**, 1395 (1983)
3 E.I. Rashba, V.B. Timofeev: Sov. Phys. Semicond. **20(b)**, 617 (1986)
4 I.V. Kukushkin, K. von Klitzing, A. Plaut, K. Ploog, H. Buhmann, W. Joss, G. Martinez, V.B. Timofeev: Pis'ma Zh. Eksp. Teor. Fiz. **51**, 575 (1990); Phys. Rev. Lett. **65**, 1050 (1990)
5 H. Buhmann, W. Joss, K. von Klitzing, I.V. Kukushkin, G. Martinez, A. Plaut, K. Ploog, V.B. Timofeev: Pis'ma Zh. Eksp. Teor. Fiz. **52**, 925 (1990); Phys. Rev. Lett. (in press)
6 C.C. Grimes, G. Adams: Phys. Rev. Lett. **42**, 795 (1979)
7 A.F. Andreev: In *Progress in Low Temperature Physics*, Vol. 8, ed. by D.F. Brewer (North-Holland, Amsterdam 1982) p. 69

From Physics to Synergetics

H. Haken

Institut für Theoretische Physik und Synergetik,
Pfaffenwaldring 57/4, D-7000 Stuttgart 80 (Vaihingen),
Fed. Rep. of Germany

Abstract. Synergetics is an interdisciplinary field of research that deals with complex systems. In particular it studies those situations in which systems change their macroscopic behavior qualitatively and are able to form macroscopic spatial, temporal, or functional structures. Its main concepts are stability and instability, order parameters and enslaving. In this contribution I demonstrate the applicability of these concepts and the corresponding mathematical methods by means of some selected problems in biology, namely morphogenesis, coordinated movements, EEG analysis, and models for recognition,including that of ambiguous patterns by humans. The latter approach is based on the concept of the synergetic computer, which represents a genuine alternative to the presently discussed neurocomputers.

1. The Problem of Complex Systems

The more science proceeds, the more it gets interested in an adequate treatment of complex systems. These are systems composed of many parts which interact with each other in a complicated fashion and may exhibit complex behavior. Examples are plants and animals, large cities, societies, or the economy. Because of the complexity of these systems, there are most probably many approaches or strategies for coping with these systems. Some of these approaches are known, many of them still to be developed. Synergetics [1], [2] may be considered as one of these approaches. As we shall see below, its strategy is as simple as general. It says: If one wishes to understand a complex system, look at those situations where it changes its behavior qualitatively on macroscopic scales. This strategy has been derived from the study of a number of processes in physics so that we shall discuss some of these processes and their theoretical treatment first.

An important step in the study of many systems is the identification of their constituents and the properties of these parts. We just have to think of the pioneering work which led to the discovery of the atom or to the establishment of quantum mechanics which allows one to explain an enormous number of properties of matter. Already at the level of one or a few atoms, an enormous mental effort was necessary to determine experi-

Springer Proceedings in Physics, Volume 57
Evolutionary Trends in the Physical Sciences Eds.: M. Suzuki and R. Kubo

mentally or theoretically their properties and the Klein-Nishina formula for scattering is a famous example of this [3]. I think it is fair to say that in biology the discovery of the genetic code in the form of DNA molecules was a similar step though the properties of a DNA molecule are of course incredibly more complex than those of a single atom. When physicists then tried to treat systems composed of several or many atoms, such as molecules or crystals, it soon turned out that entirely new ideas were needed, because these entities showed collective properties which are not present at the level of a single atom. Just think of a sound wave which is a property of a gas but not of a single atom. The situation in biology, or sociology, or economics may be similar. Take biology as an example. Animals show an incredible degree of coordination of enormous numbers of cells in breathing, blood circulation, or locomotion. This cooperation is still more pronounced when perception or thinking in humans is involved. In view of the billions of cells, it is quite obviously an unsolvable task to cope with such a problem by solving, say, the Schrödinger equation of this many-particle system, where even the biggest computer fails. Quite evidently, new concepts to cope with these systems are needed, concepts which take care of the newly emergent properties of the macroscopic system which are not present at the microscopic level.

In view of the complexity of these problems, it seems advisable to look for general methods or examples, and I wish to show that physics may show us some path through the jungle of complex systems. One of the most salient features of biological systems - but equally well of sociological and economic systems - is the fact that well-organized structures may emerge which in the case of biology come about even by self-organization.

My article will be organized as follows: First I will introduce the basic concepts of synergetics by means of the explicit example of the light source laser. Then I will give a brief outline of microscopic, macroscopic and phenomenological synergetics. Then I will show how these methods can be applied to a number of phenomena in biology, and, eventually, I will give an outlook for further applications of these concepts.

2. The Laser as a Trail-Blazer of Synergetics

The laser [4] is still my favorite example when I wish to explain the basic concepts of synergetics. As is well-known, a typical example of a laser is the gas laser, where a gas of atoms is enclosed in a glass tube whose endfaces are mirrors. They serve to reflect light running in an axial direction very often so that it can intensely interact with the atoms. The atoms may be excited by an electric current in a gas discharge. In the usual lamp, the individual atoms are excited and emit then consecutively their light independently of each other, thus producing a microscopically chaotic light field. When the current is increased and thus the atoms are excited at a sufficiently

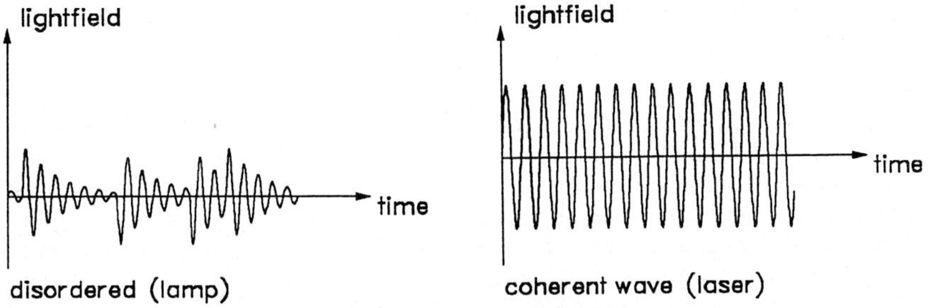

Figs. 1: The electric field strength versus time
L.h.s.: Emission from an ordinary lamp
R.h.s.: Emission from a laser

high rate, laser light is generated which consists of a practically infinitely long wave track on which only small phase and amplitude fluctuations are superimposed (Fig. 1).

In the terminology of synergetics, the emerging laser light wave serves as the order parameter which describes the macroscopic order and gives orders to the individual electrons of the atoms how to behave. The order parameter "light wave" enslaves the electrons of the atoms. On the other hand, the electrons support the existence of the laser wave so that one may speak of circular causality. In specific laser configurations, the electrons may, in principle, support several order parameters which may then either cooperate or compete. When they compete, single mode laser action may evolve. When the laser is pumped more highly, other states may occur instead of the coherent wave, e.g. ultrashort pulses or laser light chaos (Figs. 2).

The detailed mathematical treatment of these phenomena has led us to formulate a general approach to the treatment of complex systems. This approach may take place at different levels, namely at the microscopic, the macroscopic, and the phenomenological level. We shall briefly describe them in the next section.

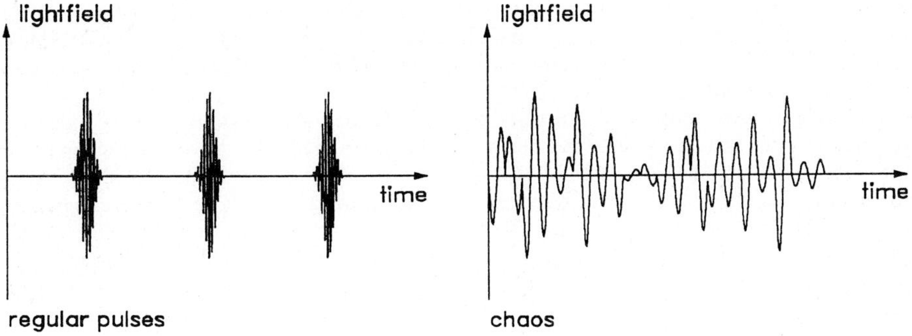

Figs. 2: L.h.s.: Ultrashort light pulses emitted from a laser
R.h.s.: Chaotic laser light

3. Microscopic, Macroscopic, and Phenomenological Synergetics

In the microscopic approach [1], [2] we start from microscopic or mesoscopic equations of motion which may be either classical or quantum-mechanical equations. In the latter case, the equations are of the Heisenberg type. In the present context we shall be concerned primarily with classical equations, however. According to them the state of the system is described by a state vector $\underset{\sim}{q}$, whose components may be for instance the laser light field, the polarization and inversion of atoms, or in fluids the density, velocity, and temperature fields, etc. The state vector undergoes a change in time which is described by a nonlinear function of the state vector and which may depend on certain control parameters α by which the system is controlled from the outside, e.g. the power pumped into the laser by the electric current in a gas discharge:

$$\dot{\underset{\sim}{q}} = \underset{\sim}{N}(\underset{\sim}{q},\alpha) + \underset{\sim}{F}(t).$$

$\underset{\sim}{F}(t)$ represents internal or external random forces. When the system changes its behavior qualitatively on a macroscopic scale, one may identify so-called order parameters by a linear stability analysis. The order parameters which are in general much fewer in number than the original degrees of freedom obey nonlinear equations which can be derived at least in principle. The individual degrees of freedom of the system are then uniquely determined by the order parameters. The enslaving principle holds not only when the order parameters describe a regular motion, such as an oscillatory (limit cycle) or a quasi periodic, but also a chaotic motion.

In many systems, such as those of biology, quite often neither the microscopic variables nor the equations of motions are known explicitly. However, two strategies have proved useful. The first one is based on macroscopic observations in which correlation functions up to the fourth order have been determined experimentally. Then via the maximum information entropy principle [5], it is possible to derive either the static distribution function for the system from which the order parameters can be identified, or one may derive an Itô Fokker-Planck equation for the time evolution of the distribution function of the observed variables [6].

A third approach which has proven most successful in a number of cases of biological systems is the so-called phenomenological synergetics where one tries to identify order parameters and then to model their time evolution by adequate nonlinear phenomenological equations. In the following we shall be concerned with the microscopic and phenomenological approach while I refer the reader to my book "Information and Self-Organization" [6] where the macroscopic approach is concerned.

4. An Approach to Morphogenesis, Finger Movements, and EEG-Analysis

During the formation of a living being, e.g. a zebra, or in other words, during its morphogenesis, the originally undifferentiated cells of the egg must differentiate. Experiments demonstrate that this differentiation depends on the position of a cell in the cell tissue so that positional information must be given to each individual cell. The idea of how this information is transferred can be traced back to the famous English mathematician Alan Turing [7] who showed that patterns can be formed by the exchange of chemical reactants. This concept has been generalized by a number of people, e.g. by Gierer and Meinhardt [8], whose model we have used as a starting point for our calculations [9]. In this model it is assumed that, in a tissue of cells, activator and inhibitor molecules are produced, may diffuse, and may interact with each other. Via diffusion and interaction specific spatial patterns are formed in which at some places the concentration of the activator molecules is high. At these places the activator molecules may switch on genes which cause the differentiation of the cell concerned. While Gierer and Meinhardt found a polarity so that the concentration of the activator is high at one end of the animal (hydra) and low at the other, thus giving an explanation of why a head is formed at one end and a foot at the other, we found that more complicated patterns may be formed as well, e.g. stripes on a zebra, or rings on the wings of butterflies. These findings were made possible by the concepts of order parameters and enslaving. But since this is older work I shall not dwell on our approach here. Rather I wish to report some more recent work, namely on finger movement and on EEG analysis.

When the American physiologist Scott Kelso asked test persons to move their fingers in a parallel fashion, it turned out that the persons had to change their movement involuntarily when they were requested to move their fingers more quickly than at a critical frequency (Fig. 3). We identified such a switch as a nonequilibrium phase transition [10], [11] which bears a close analogy, for instance, to nonequilibrium phase transitions in lasers [12]. The order parameter is the relative phase, ϕ, between the two fingers and obeys a typical order parameter equation in the form $d\phi/dt = - dV/d\phi$. The potential function

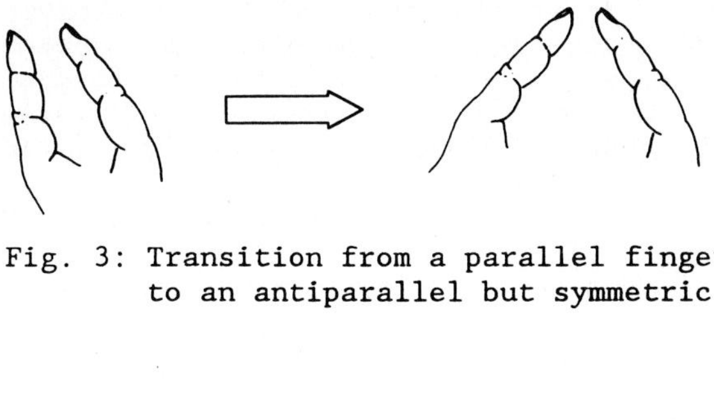

Fig. 3: Transition from a parallel finger movement (l.h.s.) to an antiparallel but symmetric finger movement (r.h.s.)

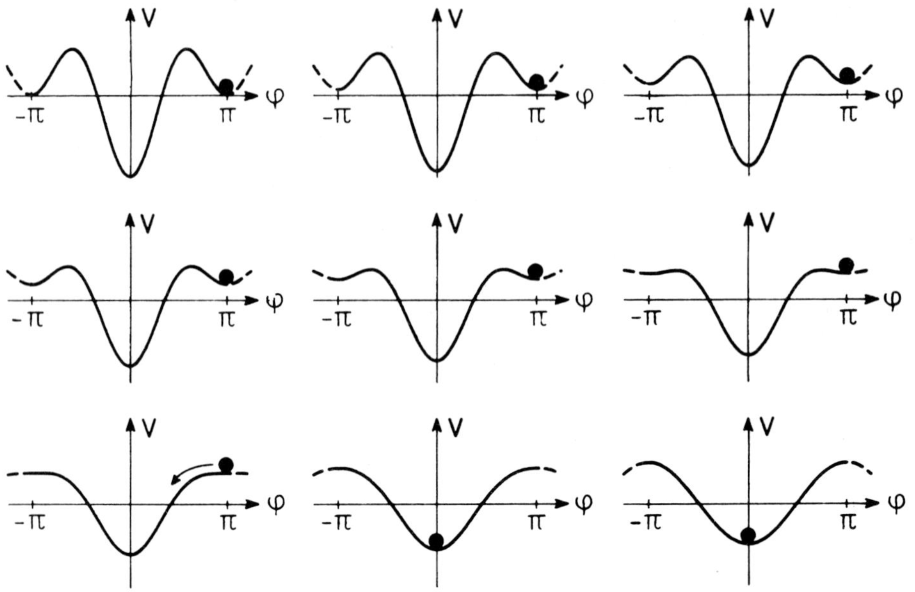

Fig. 4: The potential function of the Haken/Kelso/Bunz model [10] when the parameter b/a is changed with increasing frequency. The destabilisation of the formerly stable state at $\phi = \pi$ is clearly visible

V is plotted in Fig. 4 where b/a is a control parameter which depends on the frequency of the finger movement. Evidently the form of the potential changes so that the formerly stable state at $\phi = \pi$ becomes unstable. During the transition from the now unstable state to the lower lying stable state, typical phenomena, such as critical fluctuations and critical slowing down, were predicted and observed [11]. These phenomena are quite similar to those observed in physical systems close to non-equilibrium or equilibrium phase transitions. Subsequent work by Kelso and his coworkers [13] has shown that our concept can be applied to a variety of other phenomena of muscle coordination and gives interesting access to the problem of learning. Another example of how concepts of synergetics can be usefully applied in the case of complex systems is EEG analysis of the human brain. As is well-known, the human brain may produce electric waves in a frequency range which starts at low frequencies around one Hz and extends far beyond fifty Hz. We started our analysis of data in the α-wave region (around 10 Hz) (Fig. 5) under the assumption that the EEG patterns are produced by a system (namely the brain) which is kept close to an instability point so that the system can adapt quickly to newly changing situations. As we know from synergetics, close to instability points only few spatial modes governed by few order parameters are excited so that we attempted to study the formation of the EEG patterns from this point of view. To put our approach on a firm basis, we used the Loeve-

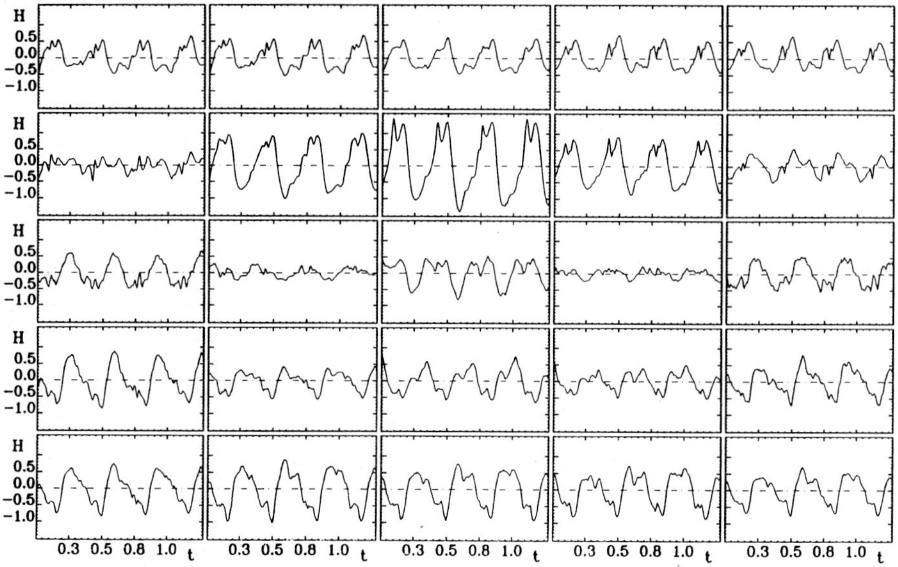

Fig. 5: Time series of an electric potential derived from electrodes at 25 positions on the scalp (after Lehmann, private communication)

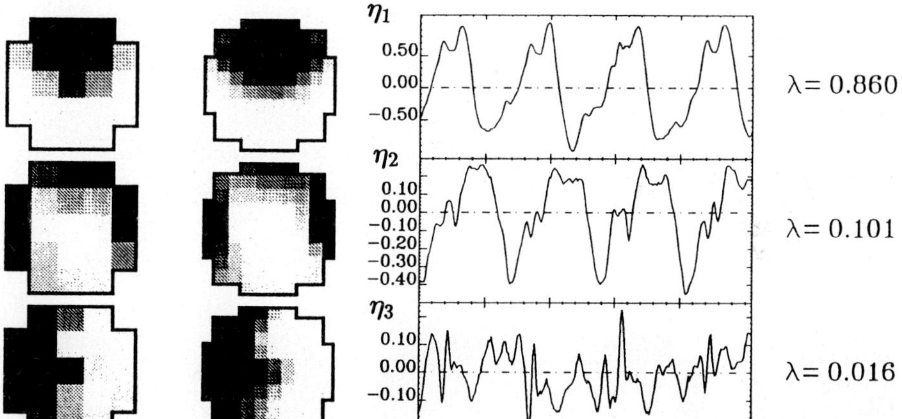

Fig. 6: The three basic modes in an epileptic seizure as derived by means of the Karhunen-Loeve expansion (after [15])

Karhunen expansion to identify the fundamental modes and it turned out that both in normal α-wave action as well as in epileptic seizures very few modes were sufficient to reproduce the spatio-temporal patterns (Fig. 6) [14]. The amplitudes of these individual basic modes are, of course, the order

Fig. 7

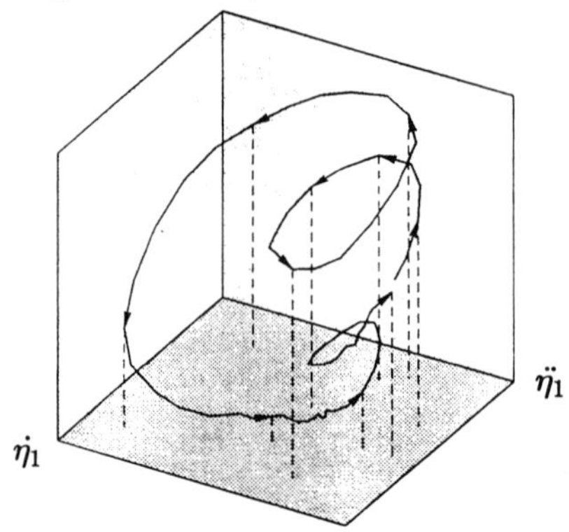

Fig. 8

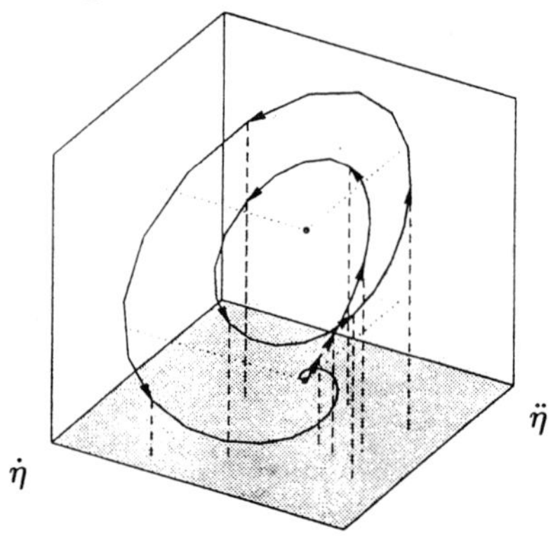

Fig. 7: The epileptic attractor derived from the time series of the electric potential in the case of an epileptic seizure (after [15])

Fig. 8: Calculation of the epileptic attractor by means of order parameter equations (after [15])

parameters. It was possible not only to recover previous results in particular by Agnes Babloyantz [16], who in her pioneering work had identified the low dimension of EEG waves and the epileptic attractor, but also the explicit order parameter equations in the case of epileptic seizures [15]. They are three coupled ordinary nonlinear differential equations (Figs. 7,8).

We could also identify the spatial modes in normal α-waves. Their spatiotemporal behavior can be represented in a surprisingly simple fashion when we plot them on a sphere (Fig. 9) [17]. It turns out that it is not sufficient to analyse the order parameters by means of chaos theory or time series analysis. As it appears, the order parameters obey for some time

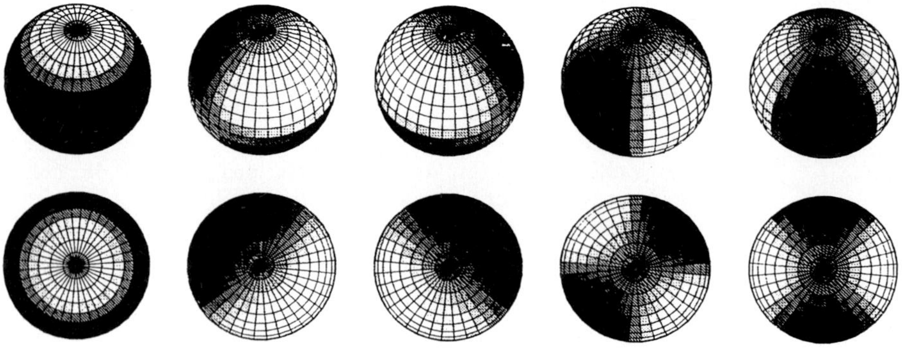

Fig. 9: An example of the presentation of α waves on a sphere (after [17])

a dynamics within a strange attractor but then are pushed to a new attractor in a rather irregular fashion.

Our findings seem to indicate a limitation to time series analyses which aim at identifying attracting states. For the time being this is a side-remark only, however.

Let us now turn to more complex phenomena which occur for instance in perception.

5. Synergetic Computer and Pattern Recognition

It is by now well-known that serial computers cannot serve as a model for human visual perception nor are they adequate machines for pattern recognition. The human brain can process visual perception at a very high speed in spite of the fact that its individual constituents, the neurones, are very slow elements. This has led scientists to the conclusion that the brain processes information in parallel and to the study of neural networks, e.g. in the form of neurocomputers. Basic ideas are due to McCulloch and Pitts [18]. This field has experienced an enormous development by the finding of Hopfield [19] that such a network can be realized, at least in principle, by a physical system, namely a spin glass. In the sense of thermodynamics, a spin glass is a closed system which possesses a specific temperature. I have found an alternative way [20] for the realization of parallel computers which is based on synergetics which deals with open systems. Synergetics has revealed general principles which apply when systems form spatial or spatio-temporal patterns. The concept of the synergetic computer is based on three ingredients:

1. The concept of associative memory which has been proposed by a number of authors, in particular by Kohonen [21].
2. The idea that pattern recognition is a dynamic process in which an order parameter or some order parameters move in a potential landscape (Fig. 10). This picture was produced in the first edition of [1], 1987.
3. Pattern recognition is nothing but pattern formation (Fig. 11) [22].

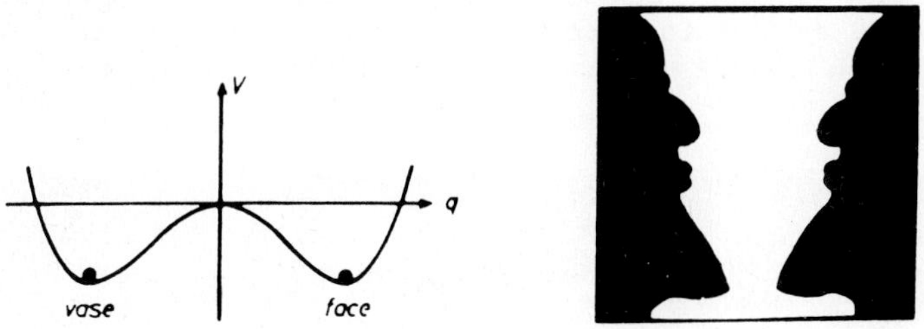

Fig. 10: Representation of pattern recognition (r.h.s. vase or faces) by means of a dynamics of an order parameter in a potential field V (l.h.s.) (after [1])

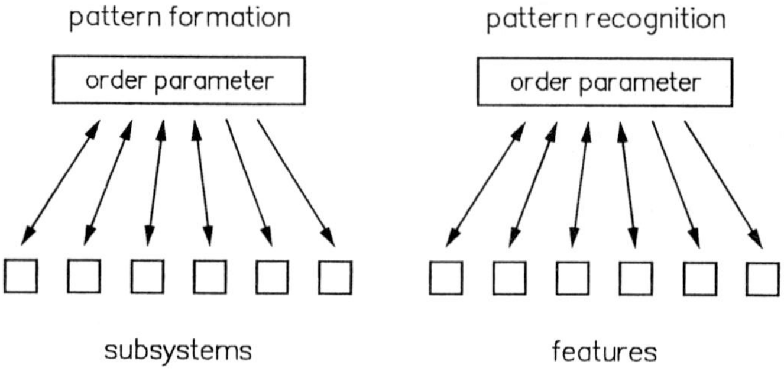

Fig. 11: Analogy between pattern formation and pattern recognition (compare text)

As we may show in synergetics, pattern formation proceeds as follows: Once part of a system is in an ordered state, it may generate its specific order parameters which may compete with other order parameters but wins the competition because of its initially given bias. The order parameter enslaves the whole system and thus establishes the totally ordered state. Similarly, in pattern recognition features of a pattern are provided which generate the order parameter. It competes with other order parameters, but then eventually wins the competition and calls up all the remaining features so that the original pattern is completed. Because this approach has been published on a number of occasions (cf. for instance [20]), I shall not dwell on the details here. Rather I want to provide the reader with a number of examples. Fig. 12 shows a number of stored prototype patterns in the form of faces jointly with letters which encode the names of the persons photographed. When part of a face is offered to the computer, it can not only complete the face but also supplement it with a letter so that the face has been recognized (Fig. 13) [23]. The procedure has been made invariant against translation, rotation, and scaling [23]. In order to identify faces within complex scenes, such as that of Fig. 14, specific control parameters in the equations describing the pattern recognition process can be put in relation to psychological attention. This has led us to devise a computer [24] which shows oscillations in the perception of ambiguous patterns, i.e. for instance in Fig. 10 a vase is recognized, then two faces, then a vase again, etc. In reality these oscillations are subject to

Fig. 12: Example of prototype patterns stored in the computer

Fig. 13: Reconstruction of a part of a face into a full face plus letter by means of the synergetic computer (after [23])

Fig. 14: A typical example of a complex scene which could be analysed by a synergetic computer (after [23])

fluctuations which we could model by means of adding fluctuating forces to the dynamics of the attention parameters [25]. Our more recent work with the aid of this synergetic computer was concerned with the identification of specific movements; for instance the computer can distinguish between a man walking or limping [26].

6. Outlook

The explicit examples mentioned above indicate that the synergetic approach which originated from physics may prove a viable concept to cope with complex systems. The confinement of this approach to situations where the macroscopic behavior of a complex system changes qualitatively is on the one hand certainly a limitation, on the other hand has allowed us to reveal general principles in the spontaneous formation of spatial, temporal, or functional structures. As we could see, in those situations the dynamics of the whole system is governed by a few degrees of freedom, the so-called order parameters. This has allowed us incidentally to establish close analogies between otherwise seemingly quite different systems. The question arises of how far these concepts may be applied to other processes in biological systems as well as to sociology or economics. In this respect some general remarks may be in order.

In our present day world, external as well as internal conditions may change rapidly. In addition, as chaos theory teaches us, systems may change internally in an unpredictable manner or in a manner which can be predicted only for rather short time intervals. Thus it will become more and more important to learn how to steer such systems. There are at least two messages from synergetics, namely:

1. Steering by indirect control via control parameters is in many cases far more efficient than a direct control.
2. In order to keep the system adaptable, we must keep it close to instability points from where it can be driven by small cues into the new state wanted.

I am convinced that such principles have been realized by nature for a long time. A recent workshop which I organized jointly with the physiologist Hans-Peter Koepchen has substantiated this suspicion. As it appears, systems like the heart, blood circulation, and breathing are kept at such points from where they can adapt very quickly and in this way develop a variety of behavioral patterns. One may speculate as to how far these concepts will carry us when we try to simulate thinking. In my interpretation, forming thoughts is again a process of pattern formation at various levels, namely at the microscopic level again and again new patterns of neuronal excitations are formed. But at the abstract level new ideas are formed. They compete with each other, they cooperate, they form new associations, and one gets the feeling that thought processes can be compared with the formation of new order parameters which enslave the subsystem of the neurones and via circular causality are caused in turn by the excitations of the neurones.

It will be a challenging task for the future to produce counter examples to the concepts of order parameters and enslaving so that limitations of these concepts will become visible and thus give ground to new development. On the other hand at the present moment I don't see such limitations and I am convinced that synergetics will find many further applications in a variety of fields.

References

[1] H. Haken, Synergetics, An Introduction, 3rd ed., Springer, Berlin (1983)

[2] H. Haken, Advanced Synergetics, Instability Hierarchies of Self-Organizing Systems and Devices. Springer (1983)

[3] O. Klein, Y. Nishina, Z. f. Physik 52, 853 (1929)

[4] H. Haken, Laser Theory, 2nd print., Springer, Berlin (1984)
H. Haken, Laser Light Dynamics, North-Holland, Amsterdam (1985)

[5] E.T. Jaynes, Phys. Rev. 106, 4, 620 (1957); Phys. Rev. 108, 171 (1957)
[6] H. Haken, Information and Self-Organization, Springer, Berlin (1988)
[7] A.M. Turing, Phil. Trans. Roy. Soc. B 237, 37 (1952)
[8] A. Gierer, M. Meinhardt, Lectures on Mathematics in the Life Sciences 7,163 (1974)
[9] H. Haken, H. Olbrich, J. Math. Biol. 6, 317 (1978)
[10] H. Haken, J.A.S. Kelso, H. Bunz, Biol. Cybern. 51, 347-356 (1985)
[11] G. Schöner, H. Haken, J.A.S. Kelso, Biol. Cybern. 53, 247 (1986)
[12] R. Graham, H. Haken, Z. Phys. 213, 420 (1968), 237, 31 (1970)
[13] G. Schöner, J.A.S. Kelso, in: Lasers and Synergetics, R. Graham and A. Wunderlin (eds.), Springer, Berlin (1987)
[14] A. Fuchs, R. Friedrich, H. Haken, D. Lehmann, Computational Systems, Natural and Artificial, H.Haken, (ed.), Springer, Berlin (1987)
[15] F. Friedrich, A. Fuchs, H. Haken, Modeling of Spatio-Temporal EEG Patterns, Proceedings Prague, Sept. 1990, to appear
[16] A. Babloyantz, A. Destexhe, Proc. Natl. Acad. Sci. USA 83, 3513-3517 (1985)
[17] A. Fuchs, Thesis, Stuttgart (1990) [18] W.S. McCulloch, W.H. Pitts, Bulletin of Mathematical Biophysics 5, 115-133 (1943)
[19] J.J. Hopfield, Proc. Natl. Acad. Sci, 79, 2554-2558 (1982)
[20] H. Haken, Synergetic Computers for Pattern Recognition and Associative Memory. In: H. Haken (ed.), Computational Systems, Natural and Artificial. Springer, Berlin, Heidelberg, New York (1987)
[21] T. Kohonen, Associative Memory and Self-Organization, 2nd ed., Springer, Berlin (1987)
[22] H. Haken, Pattern Formation and Pattern Recognition - An Attempt at a Synthesis, in: Pattern Formation by Dynamical Systems and Pattern Recognition, Springer, Berlin (1979)
[23] A. Fuchs, H. Haken, Biol. Cybern. 60, 17-22, 107-109, 476 (1988)
[24] T. Ditzinger, H. Haken, Biol. Cybern. 61, 279-287 (1989)
[25] T. Ditzinger, H. Haken, Biol. Cybern. 63, 453-456 (1990)
[26] R. Haas, A. Fuchs, H. Haken, Progress in Neur. Netw., to be published

Physics, Computation, and Biology

J.J. Hopfield

Divisions of Chemistry and Biology,
California Institute of Technology, Pasadena, CA 91125, USA

Abstract

The biological world is a physical system whose properties and behaviors seem entirely foreign to physics. The origins of this discrepancy lie in the very high information content in biological systems (the large amount of dynamically broken symmetry) and the evolutionary value placed on predicting the future (computation) in an environment which is inhomogeneous in time and in space.

1. Introduction

Why does the world of biology look so different? As physicists, we know that the workings of biology are to be explained by the known laws of quantum mechanics and statistical physics. The existence of the sun and the nature of the planet earth are important as the substrate for biology, but there are no fundamental aspects of cosmology or particle physics necessary to address the mysteries of biology. The essence of biology is fundamentally properties of molecular physics in non-equilibrium circumstances and on a large scale. By and large quantum mechanics is not relevant [1]. Of course, the quantum mechanics of chemical bonds is essential, but while the making and breaking of these bonds by the enzyme catalysts in cells is of paramount importance, it is the rates of these processes as expressed in a network of chemical reactions and not the quantum-mechanical details that matter. The essential specificity of biochemical reactions chiefly involve molecules which are so large, and binding forces which are so weak, that classical descriptions are entirely adequate. For example, the forces which hold double-helical DNA together are of quantum mechanical origin, but can be adequately modeled as effective forces acting between classical atom masses. (Contrary to the expectations of a long history of ill-prepared physicists approaching biology, there is absolutely no indication that quantum mechanics plays any significant role in biology.) So why does biology look so different to a physicist?

2. Broken Symmetry and Complexity

The first important point to note is that the micro- and macro-structures of the plants and animals which make up biology are a consequence of a massive amount of broken symmetry [2,3]. Broken symmetry, originally a part of phase transition lore in condensed matter physics, has been slowly making its way into the rest of physics. Even the laws of elementary particle physics, which would have been

Springer Proceedings in Physics, Volume 57
Evolutionary Trends in the Physical Sciences Eds.: M. Suzuki and R. Kubo

believed in 1960 to be unique, are now thought of as containing elements of broken symmetry. But in most of physics, broken symmetries are few in number.

Geology is a physical system with much broken symmetry. The fact that a particular mountain is located at a specific place and not elsewhere represents broken symmetry, as does the fact that the mountain is chiefly granite rather than sandstone. The details of a particular rock--its mineral micro composition--represent the consequence of a long and detailed evolutionary process filled with arbitrary choices. But while the minerology of a particular rock is complex, it is simple compared to that of an equivalent size piece of biological matter. This complexity can be specified by describing the set of instructions necessary to make an equivalent piece of rock. Such instructions might state: break crystals of quartz, feldspar, and alumina into small pieces, mix together, and heat at 1200 degrees Celsius for 500,000 years. A specific procedure of this kind will generate a piece of rock in every way *equivalent* to a particular geological specimen although not idential to it. They are equivalent in the sense of there being no significant macroscopic consequencences of their micro-differences, an equivalence like that of two different members of a thermodynamic ensemble. In a language appropriate for describing crystal mixtures and heat treatments for producing rocks, the program to make typical rocks need be no more than 100 bits.

If instead of describing a rock the size of a cat, you were required to specify how to make a cat from its chemical components, the problem becomes impossible. But since there are less than 1,000,000,000 bits of information in the cat genome, the description of a cat must be shorter than this in an appropriate chemical language. A genome containing 1,000,000 bits is large enough to describe a bacterium, but not a cat. So within crude limits, we know how much information is required to specify a cat in an unknown language.

When we say that biology is a complex system, it is really this immense amount of information necessary to specify the significant state of biological matter (compared to an equivalent mass of geological matter) that is being referred to. This difference is on the scale of thousands to millions, and is fundamentally involved in the apparent difference between biological systems and physical systems.

3. Stability and Evolution

Broken symmetry situations which persist for appreciable times must be stable against perturbations. The most elementary broken symmetry events are those of static stability. Statistical mechanics or thermodynamics can provide this stability, as in the case of the direction of magnetization of a ferromagnet. In the case of macroscopic systems, the laws of classical mechanics may suffice, as in the case of a block lying on the table with a particular side up. More complex cases of broken symmetry occur in open systems, in which the flow of energy through the system provides the stability to a dynamical system. Examples of this range from the great red spot of Jupiter to the mundane "flip-flop" storage circuit in digital computers. The stability of biological systems is generally a case of dynamical broken symmetry.

The ability of systems of finite size to maintain a dynamically stable broken symmetry situation is always limited. Fluctuations will inevitably cause a finite lifetime of a particular broken symmetry solution. In biology, different aspects of the broken symmetry have quite different time scales. The choice of left-handed amino acids as the building blocks of proteins must have occurred over three billion years ago. The races of man, on the other hand, are believed to have diverged more recently than 100,000 years ago. Identifiable new strains of a flu virus are created every year. The finite stability of biological dynamical systems creates the diversity seen in biology by continuing to generate new molecules, structures, and species.

The notion of evolution is also unique to biology. To be sure, rocks also evolve. Granite, under high temperature and pressure, will evolve to marble. But biological evolution is fundamentally different, due to its much greater complexity. The progression from granite to marble is a change between simple forms, of low information content. It is not necessary to seed the formation of marble with a pre-existing piece. The available space of possibilities is so small that the random fluctuations of crystal growth can spontaneously generate olivine nuclei in a short time. The evolutionary progression (under appropriate physical circumstances) is essentially inevitable. Even when there are competing forms of crystal growth, the number of more or less equivalently stable crystal structures is generally a few at most. What makes biology chiefly different is that the "crystal" whose structure is essential is a one-dimensional strand of nucleic acid. All sequences of DNA are similarly chemically stable, and there are about 10 exp(100,000,000) such sequences possible in a billion base piece of DNA. Of course, only a tiny fraction of these turn out to be biologically viable, but even so there are a huge number of possible species. The age of the earth has allowed the exploration of a negligible fraction of available species space. This is to be contrasted with geology, where the space of stable crystal forms is by comparison miniscule and fully explored.

Both biology and geology can replicate by replicating the information contained in a structure (DNA or a crystal form). For example, a crystal can be broken into two pieces and each used to seed the growth of a crystal equivalent to the starting one. DNA replication has much in common with this elementary process. Both examples involve duplicating the information contained in the broken symmetry description (see previous section) by templating on a physical structure, in one case a DNA strand, in the other a crystal surface. In the case of geology, this is not really necessary--viable crystal seeds are easy to produce by fluctuations. In biology, this process is essential, for a random sequence of DNA generated through fluctuations does not make a viable organism. Biology has thus evolved in a fashion which is qualitatively different from the evolution of a mineral. Information replication has been essential. With a system evolving through replication, nontrivial new forms are created by events which generate less than faithful replication of the genomic information.

Crystal growth far from equilibrium often results in a multiplicity of crystal forms which grow in a dynamic competition for "resources" (material to add) and may also continue to compete after growth by exchanging material. Given time and favorable kinetics, the most stable form may be capable of capturing all the available material in this dynamic competition. Biological competitions between species are different in that the whole notion of "equilibrium" and most stable or "best" is ill-defined in biology. If there is a "terrain" on which the dynamics is evolving downhill, it must also be one with an extremely rich local minimum structure. In most complicated dynamical systems, it has not been possible to isolate a Lyapunov function which is being optimized by the dynamical system.

4. Behavior

The term "behavior" in physics concerns the response of a system to a change in its environment. When we say water behaves as a liquid, we are really stating its ability to conform to the shape of a container, to shape itself into spherical droplets, to flow downhill, etc. in ways common to other liquids. Such behaviors are all consequences of the tendency of a near-equilibrium system to minimize its free energy. For such a system, the notion of behavior is most commonly a manifestation of Le Chatelier's principle.

Strongly non-equilibrium physical systems have responses which are more difficult to analyze. Occasionally they can be described by ideas such as extremal entropy production, but in general they are not unified by a simple principle. They can display complex dynamics, as in the case of turbulence. Because biological systems have much richer physical structure, they exhibit correspondingly more complex behaviors.

When the environment changes with time, it may be possible to divide the variables of a dynamic system into fast variables (with response times much faster than the time scale of environmental change), and slow variables. In such a case, an adiabatic separation can be made. The slow variables will adapt (i.e., change slowly) to changes in the environment, and the fast variables will simply see the slow variables as changing parameters. The motion of such a physical system can be termed "adaptive."

The environment in which a cell or organism finds itself in biology is normally complicated, fluctuating with significant correlation patterns both in space and in time. The organisms which can best compete in this environment will not merely have fast variables appropriately chosen for growth and reproduction in an average environment. In addition, use will be made of the adaptation of slow variables. But because these variables can only change slowly, the organism which is able to initiate adaptation in advance of an environmental fluctuation by a prediction of the future environment from the recent past and present is at a strong competitive advantage. Such predictions are useful even in a spatially homogeneous environment. For example, a yeast cell deprived of energy sources forms spores, which have a very low metabolic rate, and which generate new yeast cells when a rich nutrient broth is provided. Such behavior is very well adapted to survival in an environment where periods of plentiful food can be followed by long periods of deprivation. Forming spores itself takes an hour, and is a useful response to environment only because the correlations of nutrient circumstances have a long correlation time. The act of sporulation as a behavior can be thought of as a prediction by the organism that the deprivation will last a long time. The organism has, through evolution, learned about the nature of the correlation time in its environment. And an organism which "understands" its environment in this fashion has a major competitive advantage over one which does not.

Bacterial adaptation can be seen at the biochemical level. Bacteria raised in the presence of a single sugar make proteins which transport that sugar across the cell membrane, and make very little protein for transporting other sugars. When another kind of sugar is added, the bacterium begins to generate more protein for transporting the new sugar. The natural environment tends to be stable for a long time, followed by environmental change. The behavior is then appropriately predictive, representing the idea that any given environment tends to persist. If the sugar environment fluctuated very rapidly, this behavior would have no value.

Such prediction becomes immensely more important when the environment is spatially non-uniform on a scale larger than the size of the organism. In such a case, if the organism is able to move it can induce environmental changes by its own motions. The organism which develops movement patterns which take it into more favorable environmental circumstances has a great advantage. Such behaviors are identifiable even at the level of bacteria, which can sense and swim up concentration gradients of nutrients.

Crudely put, he who can predict the future from the present and make advantageous choices of action on the basis of that prediction will generally win in the game of evolution. Much of the history of evolution can be read as the evolution of systems to make environmental measurements, make predictions, and generate appropriate actions. This pattern has the essential aspects of a computational system, where the inputs are from environmental measurements, the outputs are the signals (chemical or electrical) which modulate the behavior, and the computation represents an appropriate generation of outputs in response to environmental signals.

The relationship between sensory inputs and behavioral outputs (or the signals which drive them) is the essential mystery of what appears to us as observers to be motivated biological behavior. The sensory input is a form of symbol, and the signals driving muscles or turning on genes can also be described as symbols. The behavioral computation done by the organism is to generate symbolic outputs appropriate to the environmental symbolic inputs. This is an example of computation in the sense that the term is generally understood in computer science. Indeed, the history of biology can be described as the evolution of symbol-manipulating systems.

5. Brain and Computation

The human brain is the most mysterious and complex of these biological computational systems. To understand its computations, we first describe a view of digital computation which moves away from purely logical descriptions, and can deal with the physical systems behind the mathematics. Our understanding of biological computation and its origins must come through studying the relation between computation and its underlying hardware, not computation as a logical structure.

The operation of a real digital computer for batch mode computation can be described as follows. A computer has N storage registers, each storing a single binary bit. The logical state of the machine at a particular time is specified by a binary vector 1001011000... of N bits. This binary state changes into a new state each clock cycle. The transition map, describing which state follows which, is implicitly built into the machine by its design. Thus, the machine can be described as a dynamical system which changes its discrete state in discrete time.

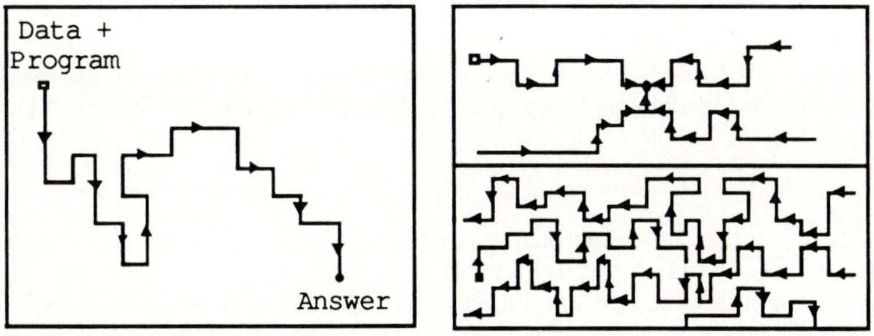

Fig. 1 The state space motion (flow) of a digital computer. In batch mode computation (left), the path goes from an initial state representing the program and data to a final stable state representing the answer. The amount of computation done depends on the complexity of the flow field. A trivial computation is shown on the upper right, and a hard one on the lower right.

The user of the machine has no control over the dynamics, the state transition flow map. His program, data, and a standard initialization procedure describe the starting state of the machine. The computation is carried out by the motion of the dynamical system. In batch mode computation, the answer represents a stable point of the discrete dynamical system, where the state space motion comes to a halt. The amount of computation which is done in this process depends on the complexity of the flow map. If it is very simple (upper right), methods can be found to locate the terminal point without following each step along the pathway. If on the other hand the dynamics is very rich (lower right), it will be essential to follow each step in order to find the answer, and the amount of computation done is then much larger.

The electrical and chemical activity of a set of nerve cells also form a dynamical system, one which moves in continuous time and with continuous state variables. But batch-mode computation can still be described in exactly the same fashion as in the digital case, as a motion to a stable attractor. (The problem of extending computation to systems with time-varying inputs is the same for digital and analog systems.) The only additional complication is the necessity of restoration. In the analog system, noise and imprecision in manufacture lead to errors in the desired trajectory. It is essential that the system recover from such fluctuations back toward the correct path in order that the computation reach the correct answer. This process is called restoration, and is unnecessary in a digital system, where operation can be made essentially perfect. The restoration process is represented by a flow pattern which locally focuses motion back onto pathways. Most of the time, states which are close to each other must lead to later states which are also close to each other. This point limits the complexity of appropriate state space motions for an analog computer. It will not be possible to use the rich complexities of chaotic dynamics in a profound fashion in biological computation.

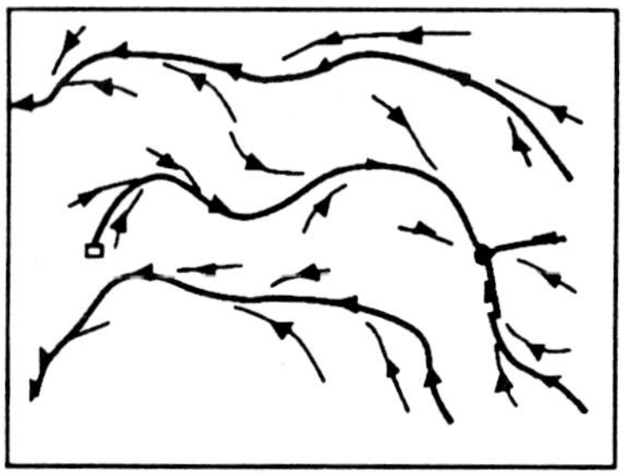

Fig. 2. The state space flow field for a batch mode analog computational system. The flow must be focussed onto paths to restore effects of errors, but is otherwise similar to those in Fig. 1.

The theoretical view of neurobiology in greatest use today represents neurodynamics as a set of first-order dynamical equations. A neuron i is generally taken as an input-output device, with output V_i given in terms of input u_i by a function such as $V_i = 1/(1 + \exp{-u_i})$. The vector u (or V) represents the state vector of activity of the system. The influence of the synapse (connection) from cell j to cell i (if any) is represented by a connection strength matrix T. The equation of motion of the activity vector is often taken to be

$$du_i/dt = -u_i/\text{tfast} + \sum T_{ij}{*}V_j + I_i \qquad (1)$$

This equation generates computation through producing a dynamics similar to those illustrated above [4].

However, the system is also adaptive. The connection strengths themselves change with time, though typically on a slower time scale. The general structure of the change with time is often represented by an equation in the style of

$$dT_{ij}/dt = -T_{ij}/tslow + del*Vi*Vj \qquad (2)$$

and involves the activity state of the neurons. These two equations are the essence of an adaptive computing system. (It is, of course, an oversimplification to represent the two time scales as completely non-overlapping). While these equations are a mere parody of the complexities of neurobiology, they contain enough of the general neurobiological themes that these equations are capable of powerful computation. There are many successful applications of such equations to real-world problems. Learning systems [5,6] have generally emphasized the computations done by an adaptive process such as Eq.2, (using Eq.1 only in a computationally trivial fashion like that on the right-hand top of Fig.1). Optimization approaches have emphasized the computation done by Eq.1, and replace Eq.2 by a set of connections provided by design [7,8].

6. Discussion

Why does biology look so different? As a physical system, it is merely another example of dynamical broken symmetry. But what sets it apart so much from other such systems is its complexity--its meaningful information content. Meaningful content, as distinct from noise entropy, can be distinguished by the fact that a change in a meaningful bit will have an effect on macroscopic behavior of a system. In addition, the meaningful bits describing the macroscopic broken symmetry are represented at the DNA level by a single long string of nucleic acids. Changing one of them can easily generate a macroscopic effect on behavior or viability. Physics is unaccustomed to looking at systems whose macroscopic dynamic properties are so influenced by single events at a molecular level, and with such massive quantities of significant broken symmetry information.

The possibility of evolution leading to a selection among information systems necessitates the transfer of a great deal of information when a structure is replicating. If this takes place through chemical templating, as it does both when a fragment of a crystal structure is used to grow a new one and when DNA is replicated, then the dimension of the informational structure must be one or two dimensional. The larger three-dimensional structure of an organism thus reflects the information carried by a tiny fraction of its matter.

The second point which sets it apart is the selection, through evolutionary pressures, of a computational system as its stable structure. It is difficult to ascertain what went on in the earliest era of the creation of biology, when the amount of dynamical information per cubic micron went from the few bits typical of physical systems to the thousands of bits essential to an elementary biological system [9]. But once a situation was established that spontaneous fluctuations could no longer generate adequately competitive forms, then the evolution of ever richer computational systems, better able to predict and to learn from the environment, was an inevitable consequence of the competition between organisms in a fluctuating but somewhat predictable environment.

7. References

[1] J. J. Hopfield in *The Lesson of Quantum Theory*, edited by J. de Boer, E. Dal, and O. Ulfbeck (North Holland, Elsevier, Amsterdam, 1986) p. 295.
[2] P. W. Anderson, Science **177**, 393 (1972).
[3] R. G. Palmer, Adv. Physics **31**, 669 (1982).
[4] J. J. Hopfield, Proc. Nat. Acad. Sci. (USA) **81**, 3088 (1984).
[5] T. J. Sejnowski and C. M. Rosenberg, Complex Systems **1**, 145 (1987).
[6] Y. LeCun, B. Boser, J. S. Denker, D. Henderson, R. E. Howard, W. Hubbard, and L. D. Jackel, Neural Comp. **1**, 541 (1989).
[7] J. J. Hopfield and D. W. Tank, Biol. Cybern. **52**, 141 (1985).
[8] Y. Takefuji and K-C. Lee, Science **245**, 1221, (1989)
[9] M. Eigen and P. Schuster, Naturwissenschaften **64**, 541 (1977); **65**, 7 (1978); **65**, 341 (1977).

Order in Molecular Biology

A. Klug

MRC Laboratory of Molecular Biology, Hills Road,
Cambridge CB2 2QH, UK

Summary

Within a living cell there take place a large number and variety of biochemical processes, almost all of which involve large molecules, particularly proteins and nucleic acids. These macromolecules often interact to form ordered aggregates or specific complexes. A number of examples are discussed which show how different kinds of order develop on grounds of geometrical or physical necessity or for reasons of functional efficiency. Examples are taken from the structure and assembly of simple viruses and the higher order organisation of the DNA double helix in chromosomes.

1. Introduction

One of the many differences between biology and the physical sciences lies in the uniqueness of biological entities and the fact that these are the products of a long history. Max Delbrück has described [1] how a mature physicist acquainting himself for the first time with the problems of biology might be puzzled by the fact that there are no absolute phenomena in biology. Everything is time-bound and space-bound.

Another outstanding feature of all organisms is their well nigh unlimited structural and dynamical complexity. Every biological system is so involved in multiple interactions and pathways, so rich in feedback devices, that one wonders whether a complete description is possible. As one goes to higher levels of organisation not all the properties of the new entity are predictable consequences of the properties of the components, no more than chemistry is, in practice, predictable from physics, whatever is, in principle, contained in the Schrödinger equation.

However, in considering the complex structures one finds in biology, one must make the old distinction, going back to Plato's 'Timaeus', between contingency and necessity. In the inorganic world this distinction may be illustrated by the problem of the shape of snow crystals, a problem which exercised Kepler and which he discussed in his book 'De Nive Sexangula' [2]. Descartes, who had read this, explained the hexagonal form of the crystals as produced by the close packing in a plane of spherical water globules and we now know that this is, in essence, correct - the internal structure involves puckered hexagonal layers of molecules. The hexagonal symmetry is thus a <u>necessity</u> and follows from what Kepler called the

Springer Proceedings in Physics, Volume 57
Evolutionary Trends in the Physical Sciences Eds.: M. Suzuki and R. Kubo

demands of matter. But what of the external shape? Many different individual shapes are found and each is contingent on the particular history of its formation. How the symmetry of the external shape is maintained during growth remains an unsolved problem.

In the living world we might also ask the same question. Does a starfish have to have a 5-fold symmetric shape? Since one can imagine a 6-fold starfish, I would say it is not a necessity, but was arrived at early in evolution which fixed the form. Does a spherical shell of a virus particle have to have, as it does, 5-fold symmetry? Here I would say yes, since, as I shall try to illustrate, it is a necessity once such shells are built as self-assembling objects constructed out of identical units. Thus the subject of my lecture is really that of order in biology on the molecular scale, and how much of it is imposed by physical necessity, purposeful as it may appear.

In the natural world around us, plants and animals often exhibit symmetry in their external forms. This geometrical regularity of living things has always been a source of wonder which over the centuries has continued to excite speculation about the invisible forces that guide the development of living organisms. From the 'Facultas Formatrix' of Johann Kepler in the early 16th century to the 'Science of Form' of D'Arcy Thompson in the early twentieth century [3], there have been attempts to make general theories but these have always foundered for lack of detail or in unfruitful abstractions. Since then embryologists have sought for more objective explanations of the patterns on which living things develop. There is a good deal of progress here, but we are really only at the beginning of finding out how the genetic program is written and implemented to tell cells how to differentiate and where to go in a whole organism.

In this lecture, therefore, I am going to limit myself to what, in another context, Galileo called the lesser phenomena - the kind of restriction which has in the history of science often proved more fruitful than grander questions about the universe at large. I shall deal with systems where the units of action are not cells but large biological molecules, macromolecules, that is, proteins or nucleic acids which form complexes or assemblies, and make patterns about which one can get precise information, by methods developed over the last thirty or forty years. This is not to say that the patterns will all turn out to be regular in a geometrical sense, though indeed some are. As one proceeds to higher scales of structure, higher levels of organisation, perfect regularity tends to get lost, but the assemblies are nevertheless ordered, according to rules we can try to grasp.

I shall use specific examples, almost all from problems I have been involved in over the years, and try to draw a more general picture out of these.

2. Specificity and Self-assembly

Many large biological structures, such as the protein coats of virus particles, filaments of muscle, or the microtubules of nerve cells, consist of ordered arrays of protein subunits, but the number of different types of units is, in general, small. Some of these can be dissociated into their components, and these can in turn be reassembled *in vitro* under appropriate conditions to produce structures which are the same as, or very similar to, the original without the need of external instructions.

These large organised structures are formed by making use of the relatively weak, but specific, non-covalent interactions between molecules. Specificity is the key word here. An interaction between biological molecules must be highly selective in that the components must recognise each other to the exclusion of other potential partners. The typical energy of a "bond", that is a contact between two such macromolecular surfaces, is of the order of 3-10 kcal (*ie* 5 to 15 kT), but the strength of the whole structure comes from the fact that there are many interactions in the assembly, and these are often cooperative. Moreover, I should stress that many of these assemblies (as I shall illustrate later) are polymorphic, and this polymorphism is vital to their dynamical role. This happens because the conformation of a protein molecule can change subtly but significantly during interactions. As a physicist would say, a protein molecule has many internal degrees of freedom and these are harnessed to develop changing interactions with other molecules of the same or different type. (It is, for example, the changes in the structure of the haemoglobin molecule when it binds oxygen that constitute the molecular aspect of breathing - it is a molecular lung, as Perutz puts it).

Thus, the most significant feature of organised structures built in this way is that their design and stability can be determined completely, or almost completely, by the specific bonding properties of their constituent units. Thus, once the component parts are made they may assemble themselves - self-assemble - without a template or other specific external control, although in the case of certain more complex assemblies, such as T4 bacteriophage, various switches in state are necessary before the assembly can proceed to the next stage.

A biological advantage of a self-assembly design for any large structure is that it can be completed specified by the genetic information required to direct the synthesis of the component molecules. The one-dimensional message contained in the DNA or RNA specifies the linear sequence of the protein for which it codes, and this in turn specifies how it will fold into a 3D structure and consequently the shape and bonding properties of the protein. Economical use of the genetic information carried by the nucleic acid of a gene will require that identical copies of some basic molecule or group of molecules will be used to build any large structure. These large structures built of subunits can also be built efficiently and with great acuracy because of the possibility of checking; *ie* any bad or wrong copy can be rejected during the assembly process.

These general ideas just mentioned derive largely from investigations begun in the late 1950's on the molecular structure of simple viruses. In the electron microscope, the virus particles are seen to have surprisingly simple shapes, cylindrical rods or spheres (really polyhedra). Biochemically they consist of long nucleic acid molecules, which carry the genetic information, encased in a shell of protein, a miniature parallel of spermatozoa.

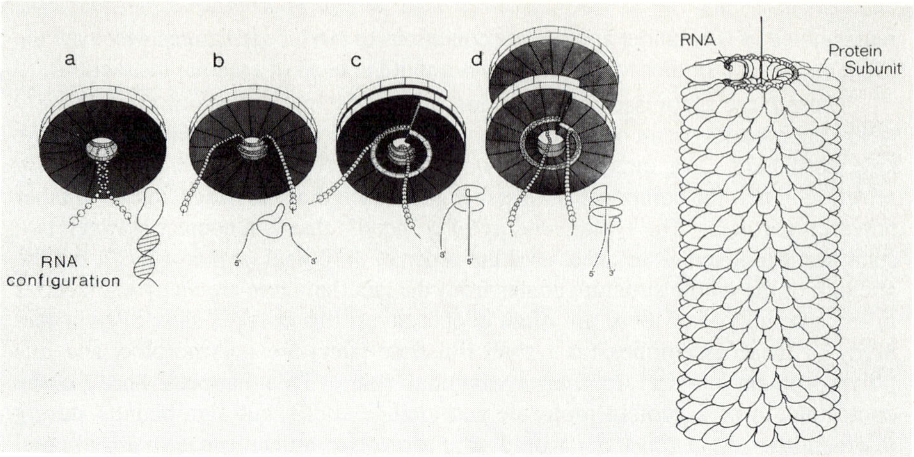

Figure 1. Nucleation of the assembly of tobacco mosaic virus (at right) begins with: (a) the insertion of the hairpin loop formed by the initiation region of the viral RNA into the central hole of the protein disk; (b) the loop intercalates between the two layers of subunits and binds around the first turn of the disk, opening up its base-paired stem as it does so; (c) some feature of the interaction causes the disk to dislocate into the helical lock-washer form; (d) this structural transformation closes the jaws made by the rings of subunits, trapping the viral RNA inside. The lock-washer-RNA complex provides the start of the helix. Additional disks then add rapidly to the nucleating complex. The special configuration of the RNA generated during the initial stages is perpetuated as the rod grows, by pulling further lengths of RNA up through the central hole, and the helix elongates to a minimum stable length.

3. The Design and Assembly of Tobacco Mosaic Virus

The classical example of a rod-shaped virus is tobacco mosaic virus (TMV) whose rod shape results from its basic design, namely a regular helical array of identical protein subunits, in which framework is embedded a single molecule of RNA wound as a helix (Fig 1, right). It is, of course, the RNA which carries the genetic information, *ie* the capacity to instruct the host cell to make many copies of the virus. This general picture was already complete by 1958, and it seemed easy to comprehend how a structure of this type might be built out of identical subunits: the subunits might assemble themselves by repeated identical interactions, like steps in a spiral staircase, enclosing the RNA as a corkscrew-like thread as the rod extends. In other words, the assembly might be likened to growth at a screw dislocation in a crystal.

We now know that this simple picture of assembly is wrong in all essentials. The virus assembles in a much more complex way, for what, in hindsight, we can see to be good physical and biological reasons. The story of how my colleagues and I came to suspect that the simple scheme was deficient and how the path of assembly was found has been told elsewhere [4] and I can only summarise the results here.

When the RNA and the protein of the virus are taken apart, the protein molecules alone under physiological conditions aggregate, not into a long helix, but into the "disk" (Fig 1a) - a two-layer cylindrical structure, each layer consisting of a ring of 17 molecules, compared with the 16 1/3 molecules present in each turn of the assembled helix. The disk can be crystallised, but because of the large molecular weight of the disk (600,000) the determination of its exact structure by X-ray methods posed formidable technical and analytical problems. These were overcome and after a dozen years it was possible to construct an atomic model showing the detailed structure of the protein subunit and how it interacts with its neighbours.

This study was pursued to the end, not merely for the sake of knowing the protein structure, but because we had earlier shown that the protein disk plays a crucial role in the assembly of the virus from its RNA and protein. The disk combines with a specific initiation tract on the single stranded viral RNA, and then dislocates to begin helix growth. Foreign RNAs which do not have this tract are rejected. When the sequence of bases of this initiation tract was determined it was seen that it could be folded so that the initial binding site is exposed at the apex of a "hairpin" structure (Fig 1a). Now the X-ray studies had shown that the two layers of the disk are so arranged as to leave a gap between them at the central hole of the disk, rather like a pair of "jaws" waiting to "bite" the RNA. So we were able to develop the picture of initiation, or nucleation as the physicist would say, of assembly shown in Figure 1. The RNA hairpin loop inserts through the central hole of the disk, and the stem of the loop opens up and binds in the "jaws" formed by the two layers of protein. The disk then dislocates into a helical structure, entrapping the RNA, after which elongation proceeds by the addition of further disks, pulling up more RNA through the central hole. There is still some controversy surrounding this picture of growth after the initial stages, but there is no doubt about the role of the disk in initiating assembly.

The disk is thus an obligatory intermediate in the assembly of the virus which simultaneously fulfils both the physical requirement for nucleating the growth of the helical particle and the biological requirement for specific recognition of the viral RNA. A most intricate structural mechanism has been evolved to give the process an efficiency and purposefulness, whose basis we now understand. TMV is self-assembling, self-nucleating and self-checking.

The general conclusion derived from the story of TMV assembly is that one must distinguish between the design of a structure and the construction process used to achieve it. In the TMV structure all protein subunits (except the few at the ends of the particle) make the same non-covalent contacts with each other, and this specific bonding pattern repeated many times leads to a symmetrical final structure. There is nothing in the design of the completed structure which gives a hint that different bonding patterns, and non-equivalent ones at that, are required during the process of assembly. This is unlike the case of the spherical viruses, where the design itself calls for departures from the precise identity of subunit packing. We turn to this next.

4. The Architecture of Spherical Viruses: Generalisations of Icosahedral Symmetry

The first attempt to understand how spherical viruses are built was made by Crick and Watson in 1956 [5]. They argued that if identical units arranged themselves regularly (or equivalently, as crystallographers say), as they do in TMV, but are designed to form a spherical rather than a cylindrical shell, then they would have to adopt the symmetry of one of the classical cubic point groups, namely tetrahedral, octahedral or icosahedral, in which cases the number of subunits would be restricted respectively to 12, 24 or 60; no regular arrangement of more than 60 is possible. The first experimental evidence from X-ray studies by Caspar on tomato bushy stunt virus, and by Finch and myself on turnip yellow mosaic virus and on polio virus, all pointed to icosahedral symmetry, a gratifying result, yet suggesting that there was somehow another special principle at work. Moreover, the first electron microscope and chemical data did not apparently agree with the X-ray ones: turnip yellow mosaic virus had 32 morphological units on its outside and contained about 150 protein units, whereas the X-ray results required 60, or a multiple of 60.

After much puzzling Caspar and I [6] were able to reconcile these apparently divergent results (Fig 2). We asked how could one build a spherical shell out of large numbers of units, abandoning the assumption that the symmetry had to be perfect. Molecular structures are, after all, not built to conform to exact mathematical concepts but rather to satisfy the condition that the system be in a minimum energy condition. Dropping the requirement of mathematical equivalence, but allowing identical units to be quasi-equivalently related, either through slight adjustments of the bonding contacts between units or through a measure of internal flexibility, we showed that the optimum design for such a shell required icosahedral symmetry, since the distortions in the specific bonds are then minimised. Certain, but not all, multiples of 60 subunits could be accommodated in such designs, which Caspar and I enumerated. Many of the hemi-spherical, "geodesic" domes designed by the architect Buckminster Fuller are built according to a similar geometry, but whereas these have to be assembled by following a fairly elaborate code, the virus shell, because of the flexibility of its units, can, so to speak, build itself.

In the 1960s John Finch and I and some of our colleagues, using electron microscopy combined with X-ray diffraction, verified that many spherical viruses - indeed every one we investigated - have their subunits arranged according to one or other of the predicted patterns, a result which holds to the present day [7]. It turns out that viruses with apparently quite different external morphologies (*eg* Figs 3b and c) belong to the same structural class. Since the construction of the icosahedron in the cube is said to be the crown of Greek geometry (it is the last theorem in Euclid), I like to think it would have delighted Plato to know that fundamental forms lay beneath the variety of appearances. To quote F M Cornford [8] on Pythagoreanism, "the key to intelligible order lies in the notion of limited quantity defining unlimited quantity, as the key to harmony lies in a few definite intervals marked out in the indefinite range of sound".

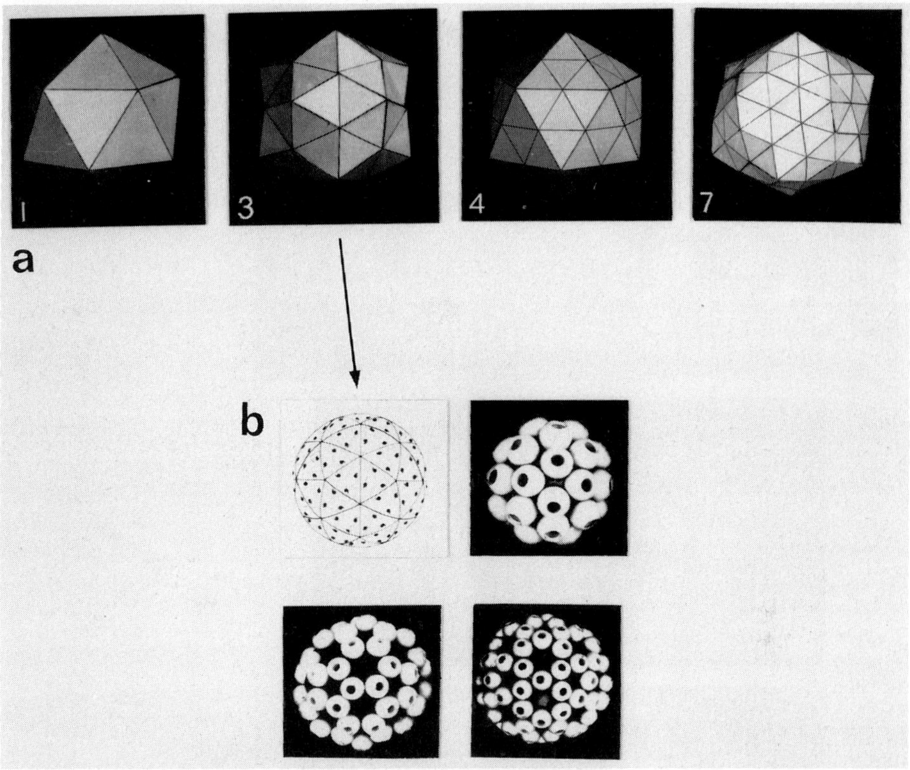

Figure 2. (a) Icosahedral surface lattices, labelled by their T numbers [6]. (b) The T=3 icosahedral lattice and its possible clustering patterns. Reading clockwise from top left are shown: the lattice showing the 3 x 60 = 180 "molecules" in general positions; the patterns of morphological units produced by clustering the molecules in 12 pentamers and 20 hexamers; in 60 trimers; and in 90 dimers. The photographs are of models and show only one side of the "virus" particle viewed down a 2-fold axis.

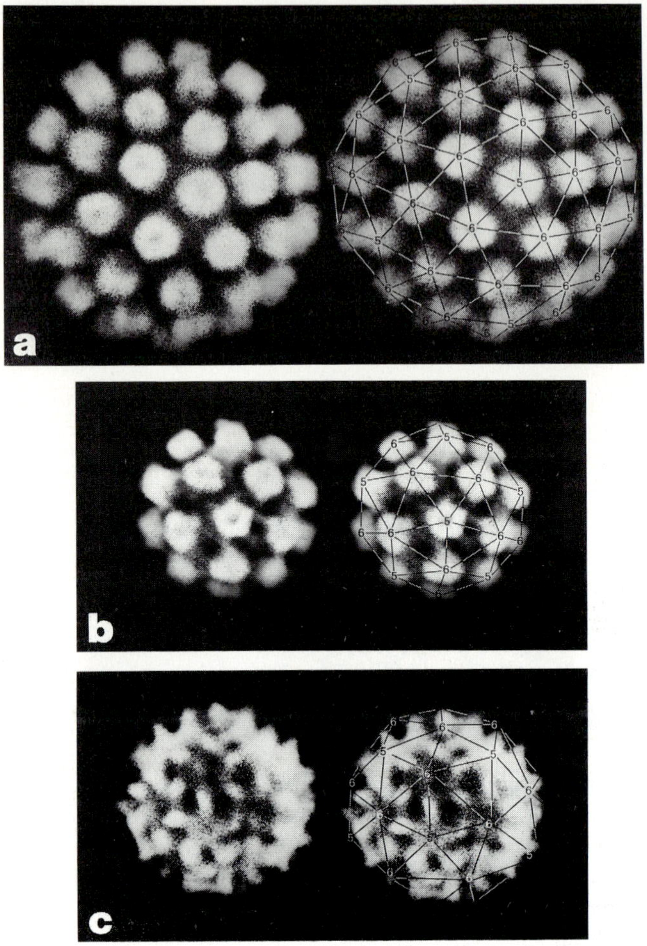

Figure 3. Three-dimensional image reconstructions from electron micrographs of some spherical viruses. Alongside are shown the underlying icosahedral surface lattices, with the 5-fold and 6-fold vertexes marked. (a) Human wart virus (about 550 Å in diameter; class T=7); (b) Turnip yellow mosaic virus (about 300 Å in diameter); (c) Tomato bushy stunt virus. (b) and (c) both belong to the same class, T=3, which has 180 units, organised around 12 strict 5-fold axes and 20 local 6-fold axes. However, the units are clustered at the surface quite differently in the two cases (cf Fig 2b): in (b) they are grouped into pentamers and hexamers around the 5-fold and 6-fold positions to form 32 morphological units, whereas in (c) they are clustered into 90 dimers about 2-fold positions.

5. Chromatin: An Ordered Hierachy of Foldings

The work on viruses has given results not only of intrinsic interest, but has also influenced general ideas on biological structure by focussing attention on the interactions between large macromolecules. Equally important, however, but not directly relevant here, the difficulties in tackling such large assemblies have led to the development of methods and techniques which could be applied to other systems besides viruses. Electron microscopy combined with three-dimensional image reconstruction, supplemented wherever possible by X-ray studies on wet materials, has provided what are now generally accepted models of the structural organisation of a large number of biological systems, such as haemocyanin, microtubules, filaments of muscle, and sickle cell haemoglobin, to name just a few diverse applications [9].

A more recent example of this approach is that of chromatin. Chromatin is the name given to the chromosomal material when extracted from the nucleus of a cell. It consists mainly of DNA tightly associated with an equal weight of a small set of rather basic proteins called histones. We took up the study of chromatin in Cambridge in the early 1970s, when protein chemists had shown that there were only five main types of histones.

The DNA in a chromosome of an animal or plant is probably a single molecule, stretching several centimetres if laid out straight. It must be highly folded to make the compact structure one sees in a chromosome (see Fig 4). At the same time DNA is organised into separate functional units - the genes. All cells, whatever their type, contain the same total DNA complement of the organism but derive their special character by expressing certain genes and not others. The major problem to understand is what controls whether a particular gene is transcribed into RNA, enabling the cell ultimately to synthesise the protein product encoded in it. The way the chromatin is folded influences such "gene expression".

We have, therefore, sought to discover the structure of chromatin. When Roger Kornberg came to Cambridge in 1972, we began using X-ray diffraction to analyse the relationships between histones and DNA. These X-ray studies showed that the native structure could readily re-form if the four histones were kept together in pairs, but not once they had been taken apart. This, together with chemical studies on the histones, eventually led Kornberg to the discovery that chromatin consists of a succession of structural subunits, called nucleosomes, containing two each of the four main histones, combined with 200 base-pairs of DNA [10]. We were later able to crystallise and solve the structure of a form of the nucleosome called a core particle, which had been trimmed with an enzyme. This crystallisation showed that almost all the DNA in the nucleus is organised at a fine structural level in a highly regular manner.

Over the years the work has continued in two main directions. We have worked down, to examine the internal structure of the nucleosome [11], and worked up, to try to understand how the filament of a nucleosome is further folded in the nucleus, giving chromatin its next higher order of structure [12].

The outcome of a wide range of structural and chemical studies is a model of the nucleosome in which the DNA double helix is wrapped in two superhelical

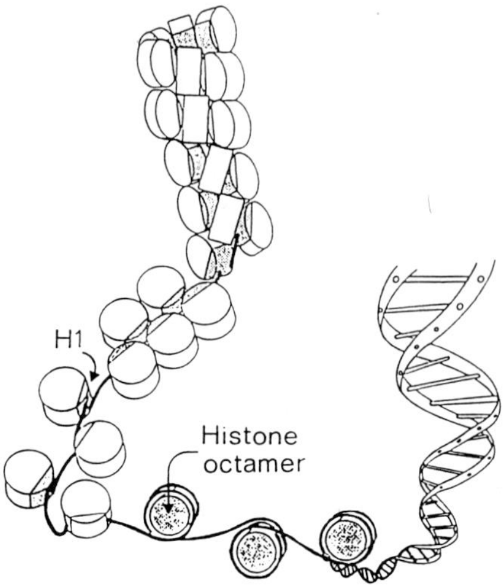

Figure 4. Schematic diagram showing the first two levels of folding of the DNA double helix (at right) in chromatin. In the first level (bottom) the DNA is wound as two superhelical turns around a flat core or spool made of eight histone protein molecules, forming structural units called nucleosomes. The next level of folding is mediated by another type of histone molecule, called H1, which attaches to the exit and entry points of the DNA on the nucleosome. The H1 molecules then aggregate into a helical polymer, as shown at left, culminating in a solenoidal arrangement of nucleosomes (top left) in which the H1 polymer runs along the centre. The drawing is idealised: the solenoidal structure is ordered, not perfectly regular.

turns around a spool (or core) formed by the eight histone molecules (Fig 4.) This DNA is sealed at the point where it enters and leaves the spool by the fifth type of histone, called H1. The H1 also mediates the coiling of the nucleosome filament into a helical or "solenoidal" structure to form the next level of structure.

Indeed, the wrapping of DNA on a protein core is only the first step in a hierarchical series of foldings which eventually results in a 10,000-fold linear condensation of the DNA into the compact chromosomes seen at the metaphase stage of cell division. The way in which this happens is not understood in detail, but there is good evidence, from the work of U. Laemmli, that lengths of the solenoidal filament, containing perhaps 50-100,000 bases of DNA, are folded back to form loops "tied" at their ends by special proteins of a "scaffold" [13]. These loops might correspond to units of gene expression or transcription: when a given transcriptional unit becomes active, the solenoid in a loop would unwind concomitantly with chemical modification of the nucleosomes. This picture of loops is also compatible with the further compaction of chromatin into the condensed metaphase chromosomes seen in cell division: the "ties" could interact together, either directly or through other non-histone proteins, to give a helical array along the axis of the chromosome.

6. Active Chromatin: DNA Recognition

The hierarchy of levels described above for packaging the DNA double helix applies to bulk of the chromatin, containing the inert genes. A current question concerns the structure of regions of active chromatin containing genes poised for expression or actually being expressed. Genes are activated and begin transcribing the DNA into RNA through the binding of regulatory proteins ("transcription factors") to a control region ("promoter") of a gene, which event, as it were, switches on the enzymatic machinery. We have chosen to work on a particular gene that codes for 5S-RNA, a component of the ribosome, since the ovaries of immature frogs contain large amounts of a transcription factor for this gene. My colleagues and I have purified this protein, called TFIIIA, and have shown that it has a remarkable repeating structure [14] Each structural unit, or domain, consists of a small loop of about 30 amino acids folded around a zinc ion (Fig 5). We have called these units "DNA-binding fingers", and have further shown that each finger binds to, and thereby recognises the sequence on, about half a turn of the DNA double helix. The fingers are, so to speak, reading heads for identifying a specific control region on the DNA.

As we predicted, these DNA-binding fingers have turned up in many other regulatory proteins. TFIIIA thus represents a novel class of proteins. Their modular design offers a large number of combinatorial possibilities for specifically recognising, and combining with, many different DNA sequences. It is likely that this design evolved by repeated duplications and mutations of an ancestral gene which coded for a single stable structural protein unit which bound to DNA. Presumably it was selected during the course of evolution because of its functional efficiency.

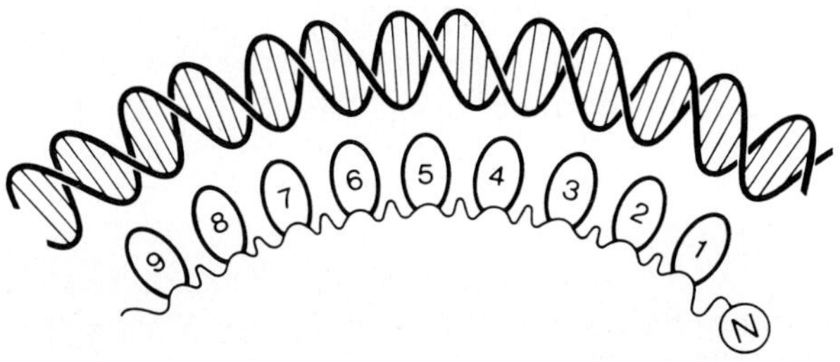

Figure 5. Schematic drawing of the structural features of the protein TFIIIA and its interaction with DNA. The bulk of the protein is organised into nine small domains, tandemly repeated. Each domain, or "zinc finger", consists of a length of about 30 amino acids folded compactly about a zinc ion, and is of a suitable size to bind into the major groove of half a double helical turn of DNA. The fingers all have a common structural framework, but derive their chemical distinctiveness from variations in a set of amino acids and residues located at the tip and on one side of the "finger". In this way the protein "reads" the varying sequence of base pairs along the DNA, and a specific interaction takes place when the two are correctly matched.

7. Concluding Remarks

With these various examples, I hope I have been able to convey to you some glimpses of how molecular biology has enabled us to understand how ordered biological systems are built, and how they function, in terms of their three-dimensional structures and of the interactions between the molecules that comprise them. We also have seen that some of the patterns produced by the forces of evolution - as in cylindrical and spherical viruses - are shaped by the same spatial restrictions that apply to mathematical objects. Others have been freely chosen to give structures and processes which have no counterpart in the macroscopic natural world.

References

1. M. Delbrück: Trans. Connecticut Acad. Sci. **38**, 173-190 (1949). {Reprinted in *Phage and the Origins of Molecular Biology* , Ed J. Cairns *et al.* (Cold Spring Harb. Quant. Biol., 1966)}

2. J. Kepler: *The Six-Cornered Snowflake*. Ed and translated by C. Hardie. (Clarendon Press, Oxford, 1966)

3. D'A. W. Thompson: *On Growth and Form*, last edition. (Cambridge University Press, Cambridge, 1942)

4. A. Klug: Harvey Lectures **74**, 141-162 (1979)

5. F. H. C. Crick, J. D. Watson: Nature **117**, 473-475 (1956)

6. D. L. D. Caspar, A. Klug: Cold Spring Harb. Symp. Quant. Biol. **27**, 1-24 (1962)

7. A. Klug: Nature **303**, 378-379 (1983)

8. F. M. Cornford: *The Unwritten Philosophy and Other Essays* (Cambridge University Press, Cambridge, 1967) p 24

9. R. A. Crowther, A. Klug: Ann. Rev. Biochem. **44**, 161-182 (1975)

10. R. D. Kornberg: Ann. Rev. Biochem. **46**, 931-954 (1977)

11. T. J. Richmond, J. T. Finch, B. Rushton, D. Rhodes, A. Klug: Nature **311**, 532-537 (1984)

12. J. Widom, A. Klug: Cell **43**, 207-213 (1985)

13. M. P. F. Marsden, U. K. Laemmli: Cell **17**, 849-858 (1979)

14. A. Klug, D. Rhodes: Trends in Biochem. Sci. **12**, 464-469 (1987)

Closing Address: History of RIKEN

M. Oda

The Institut of Physical and Chemical Research (RIKEN), Hirosawa 2-1, Wako-shi, Saitama, 351-01 Japan

As a member of the organizing committee, I am honored to close this most memorable symposium by showing a diagram (Fig.1) which briefly summarizes the history of RIKEN in the context of Nishina's life.

Japan opened the door to the western world by the Meiji Restroration in 1867. Of course Japan had had a long history of her own culture before then. But it had been totally different from European culture, which has its roots in the history of Greek philosophy, medieval Catholicism and Scholasticism. Meiji intellectuals had a great cultural shock to overcome. What they did was, first, to establish imperial universities. Secondly, in 1917, our pioneers, represented by a great businessman Eiichi Shibusawa and a great chemist who was in the United States, Johji

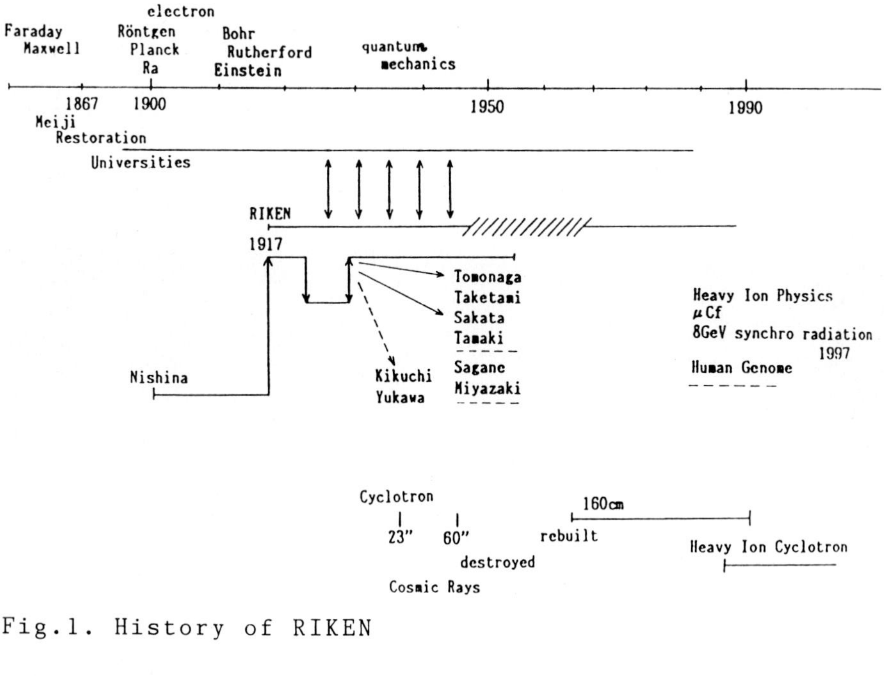

Fig.1. History of RIKEN

Springer Proceedings in Physics, Volume 57
Evolutionary Trends in the Physical Sciences Eds.: M. Suzuki and R. Kubo

Takamine, founded RIKEN with the support of the imperial family. If we read the declaration for the foundation, we note that it is pertinent even now. It reads that there are no shortcuts for the progress of industry of the nation, and the promotion of pure chemistry and physics is the only way.

RIKEN was initiated to cover a wide range of disciplines in physical, chemical and biological sciences. Nishina, as was described in the memorial session, was sent to Cambridge and then to Copenhagen in the early period of RIKEN. There, he was baptized by the "Copenhagen Spirit" and, after coming back to Japan, he manifested the spirit in RIKEN. It survived even through the most difficult 10-15 years after the War.

There is no doubt about Nishina's great personality and ability as a scientist. But what is most impressive is that Nishina's personality flourished as a result of direct contact with the European physics community in its golden age, which is represented by Niels Bohr. Now, I have noticed that due to the prominent attendees of this symposium quite a few young scientist also attending have been exposed to inspiring and active discussions. This will hopefully lead to marvelous effects in the future, extending further than Professor Kubo and the organizing committiee ever expected.

Thank you all for your active participation, which has made this symposium very successful.

Index of Contributors